81 6507227 4
TELEPEN

stems

DILEWYD O STOCK
WITHDRAWN FROM STOCK
CPC UCW LLYFRGELL LIBRARY ABERYSTWYTH
AF755072

90-220-0645-X

Simulation Monographs

Simulation Monographs is a series on computer simulation in agriculture and its supporting sciences

Simulation of grazing systems

K. R. Christian, M. Freer, J. R. Donnelly,
J. L. Davidson and J. S. Armstrong

pudoc

Wageningen
Centre for Agricultural Publishing and Documentation
1978

Books already published in this series:

C.T. de Wit and H. van Keulen. Simulation of transport processes in soils (2nd ed., revised). 1975. 109 pp. ISBN 90-220-0591-7. Price Dfl. 15.00

J. Beek and M.J. Frissel. Simulation of nitrogen behaviour in soils. 1973. 76 pp. ISBN 90-220-0440-6. Price Dfl. 12.50

C.T. de Wit and J. Goudriaan. Simulation of ecological processes. (2nd ed., revised). 1978. 180 pp. ISBN 90-220-0652-2. Price Dfl. 22.50

H.G. Fransz. The functional response to prey density in an acarine system. 1974. 149 pp. ISBN 90-220-0509-7. Price Dfl. 20.00

M.J. Frissel and P. Reiniger. Simulation of accumulation and leaching in soils. 1974. 124 pp. ISBN 90-220-0530-5. Price Dfl. 17,50

H. van Keulen. Simulation of water use and herbage growth in arid regions. 1975. 184 pp. ISBN 90-220-0557-7. Price Dfl. 27.50

G.F. Makkink And H.D.J. van Heemst. Simulation of the water balance of arable land and pastures. 1975. 87 pp. ISBN 90-220-0566-6. Price Dfl. 15.00

G.W. Arnold and C.T. de Wit (Eds.) Critical evaluation of systems analysis in ecosystems research and management. 1976. 114 pp. ISBN 90-220-0593-3. Price Dfl. 20.00

J. van den Bos and R. Rabbinge. Simulation of the fluctuations of the grey larch bud moth. 1976. 91 pp. ISBN 90-220-0589-5. Price Dfl. 16.50

R. Rabbinge. Biological control of fruit-tree red spider mite. 1976. 234 pp. ISBN 90-220-0557-7. Price Dfl. 37.50

J. Goudriaan. Crop micrometeorology: a simulation study. 1977. 257 pp. ISBN 90-220-0614-X. Price Dfl. 37.50

E. van Elderen. Heuristic strategy for scheduling farm operations. 1977. 233 pp. ISBN 90-220-0612-3. Price Dfl. 22.50

P. Kampmeijer and J.C. Zadoks. EPIMUL, a simulator of foci and epidemics in mixtures of resistant and susceptible plants, mosaics and multilines. 1977. 57 pp. ISBN 90-220-0636-0. Price Dfl. 10.00

ISBN 90-220-0645-X

© Centre for Agricultural Publishing and Documentation, Wageningen, the Netherlands, 1978.

Printed in the Netherlands

COLEG PRIFYSGOL CYMRU
THE UNIVERSITY COLLEGE OF WALES
ABERYSTWYTH
ACC. No. B78-7607
CLASS No. SB193 S6

Contents

The authors are on the staff of the Division of Plant Industry, Commonwealth Scientific and Industrial Research Organization, Canberra, Australia.

1 Introduction

One of the most obvious features of agriculture is its diversity. No two farms are alike or, as a rule, even comparable. In Australia, this diversity is accentuated by the wide range of climates and soils spanned by the agricultural region. Farm areas range from tens to tens of thousands of hectares and subdivision may be highly developed or almost non-existent. Management practices also vary widely, from the monoculture of the Merino grazier of inland New South Wales or the wheat grower of South Australia, to the intensive mixed farm of the humid coastal region.

In recent years, there has been a trend towards increased diversification within farms by the development of agricultural systems in which the production of wool, prime lamb, beef, cereal grains and oil seeds are given varying emphasis according to current economic conditions.

In a rapidly changing world, such adaptability is essential for the viability of Australian agriculture.

In contrast, the experimental worker in agricultural research attempts to reduce diversity within the system he is studying. To test a hypothesis, he compares the effects of specific treatments and, particularly in grazing experiments, most of his resources are spent in attempting to measure these effects with sufficient precision. Variability in other factors such as climate and site will tend to obscure or confuse the results and rarely will his resources allow him to test adequately their interactions with his main treatments.

Although he may regard his experiment as a physical model of a real grazing system, in fact it is unlikely to be either realistic or of general use as a model. Its realism is constrained by the inflexible control that must be imposed on management decisions if the results are to be unbiassed and analysable, its generality is limited by the many variables that cannot be controlled and by the unknown extent to which the results can be extrapolated to other seasons and sites. And yet extrapolation is essential for the practical assessment and application of the same results in real systems - systems in which management policy on grazing, conservation, drenching, marketing and so on is governed

U.C.W. ABERYSTWYTH LIBRARY

from day to day by changing circumstances.

The construction of computer simulation models offers one way of integrating experimental information into a comprehensive and quantitative description of the interacting processes involved in grazing systems. Such models may be used to assess the net impact of specific techniques on whole-farm management, and to define priorities for further experimental research.

Many models of grazing systems have already been described (van Dyne and Abramsky, 1975); these have concentrated mainly on the biological processes and consist typically of a set of functions that relate animal production to the growth and composition of pasture. As models for management they are restricted to comparing policies which require only a limited number of arbitrary decisions, most of which are made before the period of simulation starts. In this respect their structure reflects the standard experimental approach to grazing management and they might be regarded as models of grazing experiments rather than of actual grazing systems.

In practice, farm management policies are based on frequent and complex evaluations of current conditions, and not on an arbitrary and inflexible timetable. The central problem in grazing management is the dynamic allocation of resources. On one hand, the animals - the consumers - have energy demands for maintenance, on which may be superimposed varying requirements for pregnancy, lactation and growth. On the other hand, any source of animal feed may be represented as a continually changing collection of materials of different digestibilities. From this source, the animals may select according, not only to their physical capacity and energy requirements, but also to the relative importance that the manager attaches to these requirements. Although initial decisions will be made on overall strategies for matching the pattern of animal requirements to seasonal pasture production in the most economic manner, these must be reassessed repeatedly during the year to meet, for example, the conflicting needs of different types of animals at different stages of the breeding cycle and unpredictable seasonal trends.

To build this degree of realism into a model requires a more highly developed structure, in which the biological system is merely one component. Over the last nine years, we have been exploring the potential of simulation models in the application of research findings to the problems of whole-farm management. An early model (Freer et al., 1970) was designed to study the grazing of dry summer pastures, and an optimization procedure was later incorporated to evaluate set stocking and rotational grazing methods (Christian et al., 1972). However, it soon became evident that this was impractical over periods of less than a full year, because of the difficulty of assessing the residual effects of management policies on the value of body weight change and available herbage. It was also clear from this work that a grazing model should be built within as broad a framework as possible, to enable it to be adapted to a wide range of practical farm enterprises.

The structure of the model as it was subsequently developed has been briefly outlined by Christian et al., (1976). The description to be given in the following pages is designed for the use of modellers who are primarily agricultural scientists rather than programmers, and so a more detailed explanation of the program listing is provided than would be justified for a model which merely involved straightforward calculations.
A number of novel concepts are introduced in the Overview, and these require several definitions which must be clearly understood before the rationale becomes apparent. The idea is not to present a hard-and-fast procedure since, in our experience, every modeller adapts programs to suit his own purposes. What is more important is that, having grasped the functioning of these concepts in the present model, the operator should be able to adapt them to his own particular situation. We suggest that, at a first reading, it is better to gain a broad impression of the whole model as an interactive system rather than to get involved in programming details of individual subroutines. We have included, in Section 12, a guide to enable an operator to run the model with any desired combination of data and an example of the output that might be expected from such a run.

Although the biological functions that we have used here are those which appear to us to depict plant and animal performance most closely, we are well aware of the defects and deficiencies which the model contains. The measurement of system components in many cases poses almost insuperable experimental difficulties, and information concerning many of the relationships involved is at best sketchy. Compromises must constantly be made between the amount of detail which the real system contains and the limitations which are imposed not only by the capability of the computer but, more importantly, by the ability of the programmer to detect errors in logic and formulation and to comprehend the functioning and significance of each step of the whole operation. Since the model is regarded as an integral component of our research, new approaches and considerations are continually being introduced into it, and portions may be superseded. If the stage was reached at which it was decided that no further alterations were to be made it would be only because the project had been abandoned. Though seldom admitted, the process of model building is necessarily one of spasmodic and erratic evolution and improvisation; when a point of relative stability is reached, it may well indicate that the next step will result in a major upheaval. We wish to emphasize the tentative nature of this exercise since, in our experience, there is a danger that model relationships may be accepted blindly and used in situations where they may have little validity. The present listing describes the current state of the art; it is in no way the final answer.

The operation chosen for study is that of a farm producing prime lambs on improved pastures in temperate regions of Australia. It is common practice for such a property to be stocked with Border Leicester x Merino ewes, purchased at 6-12 months of age from a Merino flock in drier country. These crossbred ewes are mated at about 20 months, probably with a Border Leicester or Dorset Horn ram, to lamb sometime between late autumn and late winter. The ewes and lambs graze all the year round and obtain almost all their food from pasture, with minimal use of purchased feeds. It is increasingly common for lamb production to be combined on the one property with beef production and/or cereal cropping.

In the region used in the example, the southern tablelands of N.S.W., pasture production is severely restricted in winter by low soil temperatures and in summer by low soil moisture, and approaches its potential level for only two or three months in the spring. The farmer will usually aim to lamb his ewes late enough in the winter for the lambs to benefit from the spring growth but not so late that a high proportion of the lambs fail to reach market weight before the pasture dries off. However, many other management possibilities may modify this general pattern. The program as listed may be used to examine any of the following management factors and search for the optimum policy:

1. the degree of subdivision, relative sizes of paddocks on the farm and total area,
2. stocking rate and flock structure with respect to age groups,
3. changes in mating, weaning and selling dates,

4. set stocking as compared with various degrees of rotational grazing,
5. the extent to which animals of different ages and at different stages of the breeding cycle should be given preferential treatment with respect to their nutritional requirements,
6. the extent and timing of pasture conservation when feed is plentiful, either as deferred grazing or for hay-making,
7. supplementary feeding policy with conserved and/or purchased feed,
8. the extent to which different types of animals should be grazed together rather than as separate flocks, and
9. interactions between management policy and the levels of costs and returns.

Other problems involving practical farm decisions may be investigated by introducing suitable management factors. This model provides a general structure which will enable specific strategies to be amalgamated into a whole-farm system. Further applications which suggest themselves and which could be incorporated by direct extensions of the model include worm control programs, pasture nutrient cycling, the use of special purpose pastures, the integration of animal production with crops for grazing and grain production, and combinations of animals of different breeds or species.

In the present listing, pasture growth is generated from a seasonal pattern of ceiling yields and relative growth rates, since we consider that prediction from climatic and physiological factors cannot usefully contribute to the simulation of grazing management in the present state of knowledge. However, numerous such models have been constructed and it is probable that many research workers will prefer to use those which have been developed to meet their local conditions. This will require only the substitution of modules in two of the subroutines. In the present model, the influence of the grazing animal on the rate of herbage growth is determined solely by its effect on the amount of green material present, with no account being taken of possible changes in tiller numbers, botanical composition or similar factors. If this assumption is accepted, it is clearly more efficient to provide the model with seasonal patterns of pasture growth parameters in tabular form, rather than generate them anew for each simulation run.

The model is wholly deterministic and ignores a number of stochastic elements common to all grazing systems. Such factors might be introduced into the model, using a random number generator, in at least three areas:

a. in determining pasture growth patterns so as to simulate the unpredictability of weather changes,
b. in simulating the uncertainty associated with the incidence of metabolic diseases etc. and with changes in market conditions, and
c. in determining the distribution of body weights and other parameters within groups of animals, parameters which are at present represented only as mean values.
However at this stage it seems to us that the increase in complexity which these changes would require might obscure more important results and would add little to the accuracy of the predictions.

The program is written in FORTRAN IV and was developed on a PDP 11/10 computer with 16K words of core memory. A Control Data Cyber 76 computer is used for production runs. The operating time for one year of simulation on the two machines is 5-15 min and less than 0.5 sec respectively.

2 Overview

The model operates at three levels of organization: the *biological system, managerial control* of the biological system and *optimization* of the managerial control. A simplified diagram indicating how the various sections of the program fit into this organization is shown in Fig. 1. The coordination of these levels necessitates a more systematic approach than has been undertaken hitherto. The most important features of this approach are (a) the *classification* of the various types of animals and feed resources, combined with a systematic *coding* of the classification structures and (b) a *scheduling* routine that automatically ensures that the various program operations or *events* are carried out at the correct times. These two features are outlined below before proceeding to a description of the three levels of organization.

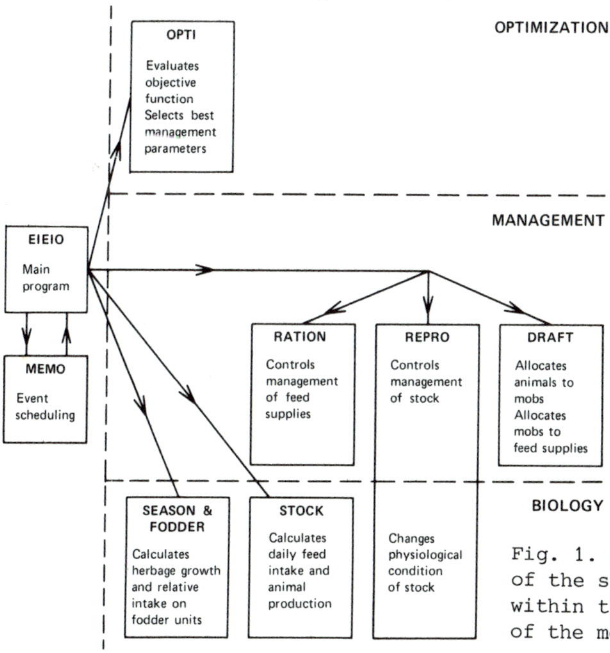

Fig. 1. Organization of the subroutines within the structure of the model.

Classification and coding

The feed supply on the farm is provided by a number of *fodder units*, while animals are grouped into *stock classes* on the basis of age or any other characteristics which make separate consideration desirable. The various attributes of fodder units and stock classes are listed in Table 1, illustrating the analogous aspects of the classifications. It should be noted that the reference number of any fodder unit (J) or stock class (K) specifies its location in the relevant arrays. For example, the area of fodder unit 6 is A(6), while the current number of animals in class 2 is S(2) and their mean body weight B(2). In the present listing, the maximum number of fodder units or stock classes is 19 but this may be increased by redimensioning the appropriate arrays.

A fodder unit may be a store of hay or grain, a paddock of pasture which is being grazed or conserved for future grazing or hay making or it may be a paddock of crop. The nature or current usage of any unit J is specified in coded form by its *status*, IJ(J), as set out in Table 10. The status of a paddock may of course be changed during the year, according to management decisions. The total feed dry matter (DM) in each unit is divided into *digestibility classes*; in the current listing there are five of these, ranging from 40% to 80%. The number of fodder units and the fixed attributes of each (size and initial amount of feed) is specified by the operator at the start of each run.
In a somewhat similar fashion, the status value, IK(K), of a stock class defines the physiological condition of the animals, as listed in Fig. 2, and it is apparent that the status of most stock classes will change several times during the year. The number of days spent at each status value, TI(I), is specified at the start of the program. Initial values of stock attributes, such as the number in each class, mean body weight and age are read in at the start of each run.
A *mob* consists of one or more stock classes which are grazed together at any given time. A mob may graze any number of paddocks simultaneously, and hence two additional variable arrays are required:
1. MK(K) specifies the mob to which stock class K belongs
2. MJ(J) specifies the mob which currently grazes fodder unit J.
An example is shown in Fig. 3. As in the practical situation, it is evident that two mobs cannot simultaneously occupy one unit, since they would then become indistinguishable. Furthermore, no class of stock can be split between two mobs, because the mean values of their attributes would then diverge. The

Table 1 Attributes of fodder units and stock classes and of the status of each

Fodder units		Stock classes	
No. of unit	J	No. of class	K
Area	A(J)	Stock no.	S(K)
DM ha^{-1} in each digestibility class L	F(L,J)	Body weight	B(K)
		Age	AGE(K)
		Wool weight	W(K)
		Body weight gain	BG(K)
Relative intake from class L	RD(L,J)	Potential intake of DM	DK(K)
		Actual intake of DOM	D(K)
		Intake of met. energy	E(K)
		Milk production	EL(K)
Time of next event	T(J,2)	Time of next event	T(K,3)
Mob which is grazing unit	MJ(J)	Mob to which class belongs	MK(K)
Status of unit	IJ(J)	Status of class	IK(K)
Attributes of unit status		**Attributes of class status**	
Priority	PJ(I)[1]	Priority	PK(I)[2]
Monetary values	VJ(I)	Monetary values	VK(I)
		Grazing pattern	FI(I)
		Time interval between status changes	TI(I)

1. I = IJ(J)
2. I = IK(K)

program assumes that diet selection and intake of all classes in the same mob are influenced in the same way by changes in feed availability and digestibility.

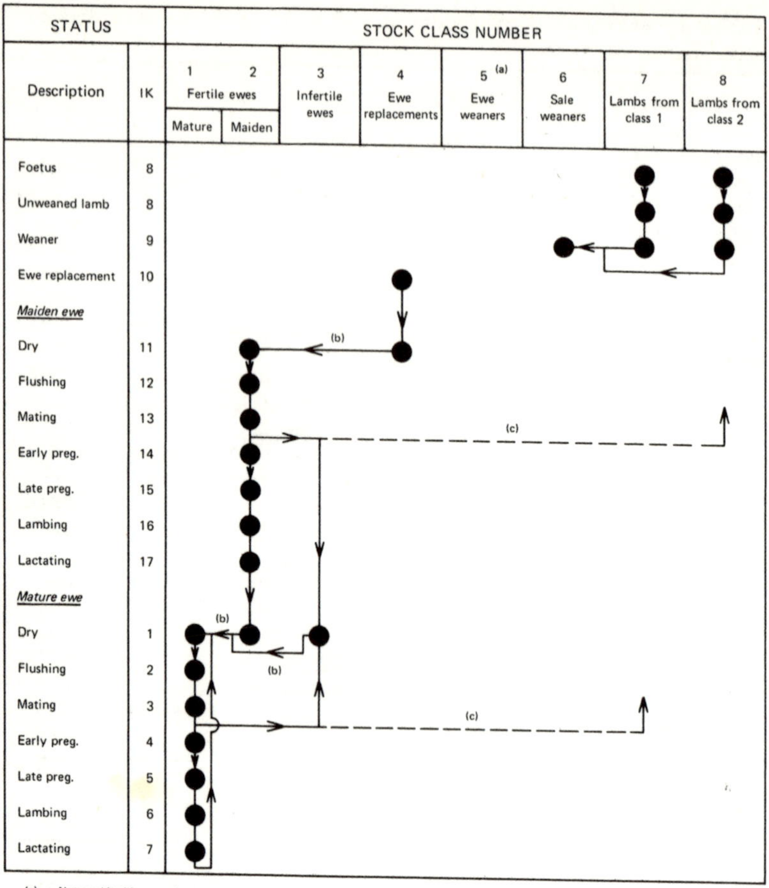

(a) Not used in this example
(b) Reorganization at the end of the year
(c) Creation of new lamb classes at mating

Fig. 2. Changes in the status of each stock class.

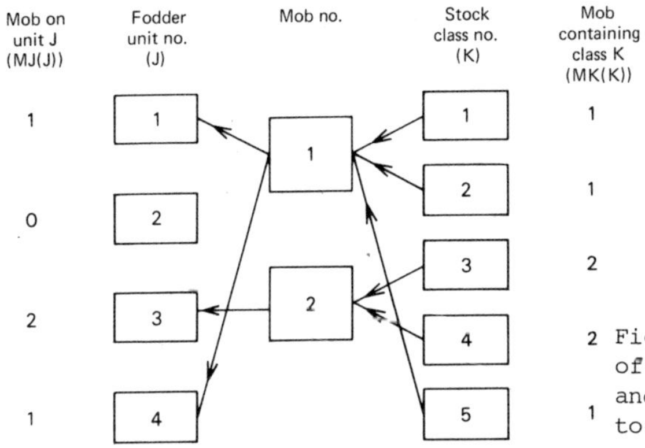

Fig. 3. Allocation of stock classes and fodder units to mobs: an example.

Scheduling of events

The simulation advances in steps by a succession of operations or events, which are maintained in chronological sequence by a list-processing procedure which forms the basis of subroutine MEMO (see Section 5). Three lists of event times are used: (a) general, (b) fodder and (c) stock. The general list in the present model contains only the following six items, though the number could be extended indefinitely to handle other farming operations:

1. daily assessments of herbage composition and animal nutrition (calls to subroutines FODDER and STOCK),
2. change in status of a fodder unit (call to subroutine RATION,
3. change in status of a stock class (call to subroutine REPRO),
4. allocation of mobs to fodder units (call to subroutine DRAFT),
5. allocation of stock classes to mobs (call to subroutine DRAFT), and
6. assessment of objective funtion and change in value of optimization parameters at the end of each run (call to subroutine OPTI).

The fodder and stock lists hold the time of the next event for each fodder unit and stock class respectively. At each step the main program takes the number of the next event from the top of the general list, advances the time accordingly and calls the appropriate subroutine. The subroutine carries out the required procedure and sets the next time for the event to occur before returning control to the main program. Subroutine

MEMO meanwhile sorts the lists to present the new time sequence in chronological order.

Biological system

At this level of the model (Fig. 1), three main processes are dealt with:
1. daily growth rates and changes in digestibility of herbage in each paddock (subroutine FODDER),
2. daily feed intakes and animal production of each stock class (subroutine STOCK), and
3. changes in physiological condition of animals, and hence in stock class status (subroutine REPRO).

Pasture growth is generated from annual patterns of ceiling yields, relative growth rates and senescence rates, as specified by the operator. A number of different seasonal patterns are read in and the operator prescribes the sequence in which they are to be used in each simulation run. These growth parameters are then used in subroutine FODDER (see Section 6) to calculate, for each paddock in turn, the daily net increase in the weight of herbage DM in each digestibility class.

Subroutine FODDER then predicts the extent to which animals grazing each paddock would be able to satisfy their appetite. Their *potential intake* is defined as the weight of DM per animal per day which would be eaten if feed availability and digestibility were non-limiting. The *relative intake* on the other hand, is a characteristic of a paddock and is defined as the proportion of the potential intake which could be obtained on that paddock. Relative intake is assumed to be the same for all types of animals and is calculated as a function of the weight of herbage DM present in each digestibility class.

Where a mob is allowed to graze more than one paddock at a time, the distribution of animals will naturally vary, but it is assumed in the model that the mean stocking rate on each paddock will be directly related to the relative intakes. Since the intake per head is also related to the relative intake, it follows that intake per hectare will be related to the square of the relative intake. This assumption is consistent with the common observation that, in these circumstances, sheep preferentially graze the better paddocks and thus gradually reduce differences in herbage availability.

Subroutine STOCK (see Section 7) then predicts daily animal performance from each stock class in turn. Potential intake is calculated as a function of age and body weight and, after allowing for intakes of milk or hay, the actual intake of digestible dry matter from pasture is obtained as the product of potential intake, relative intake and diet digestibility. Net energy balance, body weight gain, foetal development, milk yield and wool production are then predicted by conventional methods (e.g. A.R.C., 1965). Finally, the subroutine calculates the amount of material removed from each digestiblity class of herbage by the animals in each paddock.

Changes in the status of stock classes, which are made in subroutine REPRO (see Section 9), may be either biological or managerial in origin. Most of the changes in physiological condition, shown by the vertical arrows in Fig. 2, represent stages in the breeding cycle, which is initiated by the management decision to start flushing the ewes before mating.
The transitions from one status to the next then occur at fixed time intervals until weaning. The horizontal arrows in Fig. 2 represent the transfer of animals from one stock class to another, either because of an increase in age during the year or in the general reorganization of the flock at the end of the year. Functions for calculating the proportions of multiple births and infertile animals, neonatal losses and birth weights are included in the subroutine.
In the present model, the status of fodder units is not altered by biological events. However, status could be used, for instance, to denote such crop developmental stages as emergence, anthesis and grain maturation.

Managerial control

The management section of the model is responsible for the timing and execution of the following kinds of action:
1. changes in fodder unit status and other attributes, involving such operations as closing paddocks for deferred grazing, haymaking and purchase of hay (subroutine RATION),
2. decisions involving stock classes, including the start of flushing, time of weaning, sale of lambs and cull ewes and the provision of replacement ewe weaners (subroutine REPRO), and
3. allocation of stock classes to mobs and allocation of feed resources to mobs (subroutine DRAFT).
The ultimate aim of management is to obtain the most efficient matching, from an economic viewpoint, between the variable pattern of feed supply and the competing needs of growing, pregnant and lactating animals. The computer simulation of an overall management strategy which has sufficient flexibility to respond to changing pasture and animal conditions during

the year means that decisions, which in practice are made intuitively and qualitatively on the basis of past experience, must be translated into precise values and numerical procedures. The model therefore makes use of *management parameters*, the values of which determine the actions to be taken at various points in the program where decisions are called for. The purpose of many of these parameters (date for start of flushing, latest date for haymaking etc.) is immediately apparent. In other cases their significance is less obvious, and the values initially assigned to these parameters at the start of the program must be regarded as subjective estimates which may be improved upon by experience, system testing and in particular, by the process of optimization.

Among the most important of the management parameters are the *priority values* which are assigned to the fodder units and stock classes *according to their status*. These values determine to a large extent the allocation of animals to feed supplies in the procedure which is contained in subroutine DRAFT (see Section 10).
The priority value, PK(I), of any stock class K with status I (where I = IK(K)) is the minimum proportion of the potential intake that is considered necessary for these animals in relation to the productivity of the system as a whole. For example, the priority value of ewes in late pregnancy, PK(5) (see Fig. 2) might be 0.8, whereas the lower demands of dry ewes might be reflected in a priority value, PK(1), of 0.4. This priority value is raised if the body condition of the animals is below normal and reduced if it is above normal.

In contrast, the priority value PJ(I) given to a fodder unit J with status I (where I = IJ(J)) determines the extent to which feed requirement must exceed feed supply before the unit is made available for consumption, rather than being conserved. The distinction in meaning between PK(I) and PJ(I) reflects the fact that whereas animals can be provided with varying amounts of feed, fodder units in general must be either grazed or not grazed; they cannot be used partially.

The *feed requirement* for each stock class is calculated as the product of stock number, potential intake and priority value, while the *feed supply* of each fodder unit is defined as the product of area and relative intake. The allocation of mobs to fodder units is done by determining how the total feed requirement can be met from the various feed supplies, starting at the fodder unit of lowest priority, until the priority value of the next fodder unit exceeds the ratio of feed requirements to feed supplies. The fodder units with higher priority values are then ruled unavailable for consumption.

An illustration of the procedure is shown in Table 2. In the first part, the total feed requirement is calculated to be 1990. In the second part, fodder units are added in increasing order of priority and the cumulative feed supply calculated each time. When fodder units 4, 2 and 5 have been included, the available feed supply is 60, giving a feed requirement/feed supply ratio of 33. Since the priority value of the next fodder unit (3) has been set at 40, this unit and the one

Table 2 Example of the calculation of feed requirements and feed supply

Feed requirement

Stock class	Status	Priority value	Stock number	Potential intake ($kg\ day^{-1} head^{-1}$)	Feed requirement ($kg\ day^{-1}$)
6	8	0.9	300	1.0	270
4	9	0.6	500	1.0	300
5	10	0.5	500	1.0	250
3	1	0.3	2000	1.5	900
2	1	0.3	400	1.5	180
1	1	0.3	200	1.5	90
					Total 1990

Feed supply

Fodder unit	Status	Priority value	Area (ha)	Relative intake	Feed supply (ha)	Cumulative feed supply (ha)	Requirement/supply
4	1	10	20	0.6	12	12	166
2	1	10	40	0.7	28	40	50
5	1	10	50	0.4	20	60	33
3	3	40	30[1]	0.8[1]	24[1]	84	
1	4	60	–[1]	–[1]	–[1]		

1. Not applicable to hay store

= Total Req't divided by Cumulative Feed supply

remaining (unit 1) are not made available. If the feed requirement subsequently increases or the relative intakes decline so that the ratio rises above the priority value, unit 3 is opened for grazing. If the ratio rises still further, then depending on the value of PJ(4) assigned by the operator to the hay store (unit 1), hay will be fed to the grazing stock. At every change in the status of any of the fodder units or stock classes the management system is reassessed and, if necessary, a redistribution of stock classes is made in three main steps in subroutine DRAFT. Firstly, the stock classes are allocated to mobs by grouping together those whose priority values differ by less than a specified amount. Secondly, the current feed situation is assessed, in the way just described, to decide which fodder units are available for grazing. Lastly, the mobs are allocated to the available fodder units, matching feed requirements to supply as far as possible but ensuring that mobs of highest priority receive the fodder units with the highest relative intake. Whether the allocated paddocks are used by the mob in a set-stocking or rotational grazing system depends on the specific rotation prescribed by the operator for the status of the stock classes in the mob.

Optimization procedure

The third level of organization, that of optimization, has two main functions:
a. the formulation of an objective function which represents in $\$ha^{-1}$ the results of one run of the model and provides an index for comparing one run with another, and
b. a systematic search for those values of the *optimization parameters* that maximize the size of the objective function. The optimization parameters are a subset selected from the management parameters by the operator.
Both of these operations are performed by subroutine OPTI (see Section 11) which is called at the start of each run and again at the end of each year of simulation. It sums farm expenses and returns from sales and the changes in the value of feed and stock on the farm and expresses this total ($\$ha^{-1}$) as the objective function. The subroutine then directs a search for the best values of the management parameters that were selected by the operator. It does this by a step-wise process of increasing or decreasing the size of each parameter in turn until the size of the objective function increases by less than a specified increment. Although a large number of management parameters could be tested at one time, this would require an inordinate amount of computing time and one is, in practice, limited to a combination of three or four variables.

The integration of the three levels of organization is illustrated in the simplified flow chart of the whole model shown in Fig. 4.

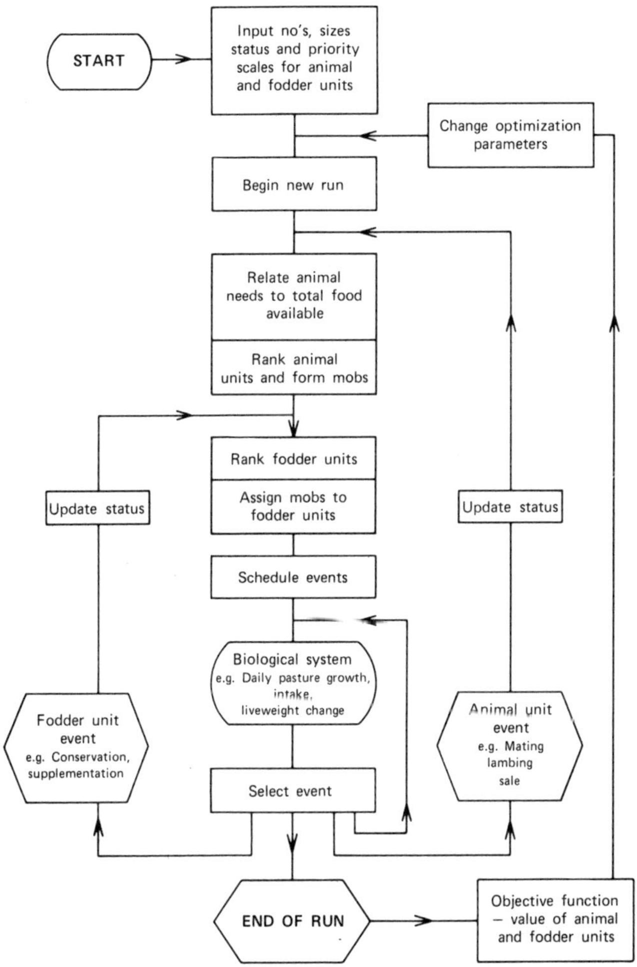

Fig. 4. Simplified flow chart of the whole model.

3 Program EIEIO

The main program is responsible for the following operations:
1. setting initial values of program variables
a. at the start of the program,
b. at the start of each simulation run, and
c. at the start of each year of a simulation run; and
2. calling subroutines in sequence during the simulation runs.

A list of the variables contained in common blocks is given in Table 3; the meanings of these are explained more fully in the relevant sections.

Start of program

A data statement (line 14) sets to zero several integer counters which are required in the optimization procedure. The value of the objective function, OOO, is given a highly negative value, ensuring that it will be exceeded on the first run. The contents of the first part of common block O, comprising various optimization variables, are also set to zero (lines 17-18).

Start of each run

The user must specify the logical unit or file number for program input and output (lines 20-21). The main data store is held on file LU, while LFG contains sets of seasonal growth patterns. Output from various parts of the program may be assigned to any number of files; eight (L1-L8) are specified in this listing. The array TLU, specifying the time intervals at which information is to be recorded on these files, is also read in (line 22), followed by the total number of runs required, NRUN (line 23). The current run number, IRUN, is incremented by one, and if NRUN is exceeded, the program terminates (lines 24-25).
Running variables in common blocks A and D are set to zero (lines 26-31), and initial values of other variables are read in (lines 32-37). Examples of the values which are required to be inserted here are given in Section 12.
Lines 38-44 are concerned with variables to be used in the optimization routine. There are NOP optimization parameters,

Table 3 Variables in common blocks

Block A	
RD(6,20)	Relative intake for each digestibility class on fodder unit
DM(4,10)	Factors for dry matter intake by mob
FM(10)	No. of steps in grazing rotation for mob
SM(10)	No. of stock in mob
QM(10)	Mean digestibility of feed eaten by mob
TT	Current time
TYR	Time of year (calendar date)
DFT	Total dry matter intake of herbage and hay
Block B	
T(20,6)	Times for next events (days) and current priorities
F(6,20)	Fodder in each digestibility class for each unit ($kg\ ha^{-1}$)
FL(5,3)	Distribution factors for digestibility class
Q(5)	Digestibility of fodder class
CFG(27,3)	Fodder growth factors for each period
FG(3)	Current growth factors
TLU(10)	Time intervals for data output on logical units (days)
Block C	
C(100)	Miscellaneous constants
A(20)	Area of fodder unit (ha)
S(20)	No. of stock in class
B(20)	Body weight of stock (kg)
BN(20)	Normal body weight (kg)
AGE(20)	Age of stock (days)
W(20)	Wool weight (kg)
DI(20)	Intake factor for stock class during lactation
EL(20)	Energy of milk production ($MJ\ day^{-1}$) (negative)
PJ(20)	Priority of fodder unit
PK(20)	Priority of stock class
VJ(20)	Value of fodder unit ($\$ha^{-1}$)
VK(20)	Value of stock class ($\$ha^{-1}$)
FI(20)	No. of steps in grazing rotation for stock status
TI(20)	Time interval between status changes for stock class (days)
TK(20)	Days from start of time interval at which next event occurs
Block D	
D(20)	Actual intake of DOM by class ($kg\ day^{-1}\ head^{-1}$)
DK(20)	Potential intake of dry matter by class

Table 3 (continued)

	(kg day^{-1} $head^{-1}$)
E(20)	Intake of metabolizable energy by class (MJ day^{-1} $head^{-1}$)
BG(20)	Body weight gain of stock class (kg day^{-1} $head^{-1}$)
MJ(20)	No. of mob on fodder unit
MK(20)	No. of mob to which stock class belongs
NT(20,6)	List of order of events and priorities in T array
Block Y	
S0(2)	Proportion of dry stock
S1(2)	Proportion of stock with single offspring
S2(2)	Proportion of stock with twins
B1(2)	Weight of a single foetus (kg)
B2(2)	Weight of a twin foetus (kg)
BO(2)	Body weight or modified body weight at a specified time
TMID(2)	Mean time of conception after start of mating (days)
Block O	
OJ(20)	Total value of fodder unit ($\$ha^{-1}$)
OK(20)	Total value of stock class ($\$ha^{-1}$)
OW(20)	Total value of wool for stock class ($\$ha^{-1}$)
OV(20)	Variable costs for stock class ($\$ha^{-1}$)
XOP(40)	Record of optimization parameter combinations tested
COP(20)	Best value so far for optimization parameter
DOP(20)	Step size in changing optimization parameter
IOP(20)	Index of optimization parameter in C array
NOP	No. of optimization parameters
NRUN	Total no. of optimization runs
NDO	Total no. of reductions in step size to be made
IYR	Year of optimization run
IO	Index in IOP of parameter currently being optimized
IRUN	No. of current optimization run
IDO	No. of reductions in step size so far
IREV	Flag denoting unsuccessful test with negative step size
IMP	Flag denoting at least one success with current step size
OO	Objective function for current run
OOO	Highest value of objective function so far
OEI	Total expenses and income for current year
OEIT	Total expenses and income for current run
Block M	
NFG(20)	Sequence of seasonal patterns for pasture growth

Table 3 (continued)

IJ(20)	Status of fodder unit
IK(20)	Status of stock class
KRE(20	New class to which this class is transferred at end of year
KY(20)	No. of lamb class which belongs to this ewe class
Block N	
NJ	Total no. of fodder units
NK	Total no. of stock classes
NL	Total no. of digestibility classes of fodder
NM	No. of mobs
NP	No. of periods in year
NDP	No. of days per period
NYR	Total no. of years in each run
LU	Logical unit for input of model parameters
LFG	Logical unit for input of seasonal patterns
L1 - L8	Logical units for output
Block G	
IIG	Dummy variable acting as IG(0)
IG(125)	Array for plotting selected parameters
LG(8)	Plotting symbols (numerals 1 to 8)
LO	Plotting symbol (blank)
LDOT	Plotting symbol (full stop)

selected from the variables in the common block C, and specified in the IOP array. Thus if C(25) is to be the first optimization parameter, IOP(1) = 25. At the start of the first run (IRUN = 1) the initial values for the optimization parameters are transferred to the COP array (lines 39 and 40) and the step sizes for changing these parameter values (DOP array) are read in (line 41). In subsequent runs, the COP values are increased or decreased by the corresponding DOP values, and the set of COP values which give the highest figure for the objective function become the management parameters used in the C array (line 44), replacing the original values read in each time in lines 34-35.

The next part of the program sets up the NT array, which has the function of listing the sequence of events and priorities in the corresponding rows of the T array, the contents of which are shown in Table 4. The procedure is described in detail in Section 5, and only the essential steps are mentioned here. The last column of each row in the NT array acts as a reference point for starting the list, and its value is originally set to 20 (line 47). Calls are then made to subroutine MEMO, which establishes complete linkages between the relevant members of each row by creating lists in which each member specifies the next in ascending numerical order of T values, with the values in the last column being set at some very large number (line 46). The first row of the NT array, referring to general events, is set up in line 49. Line 54 arranges in correct sequence the initial times of fodder events (row 2). Similarly, line 62 arranges the initial times of events (row 3) of the stock classes.
Lines 50-59 deal with fodder unit parameters. The total number of fodder units, NJ, may be optimized by selecting C(8) as an optimization parameter. The calculation which follows, in lines 51-58, enables paddock and farm sizes to be changed between optimization runs. The values of A(2)-A(19) which are read in specify proportional paddock areas, while A(20) gives the total farm area. From these figures the actual area A(J) of each paddock is obtained. The value of A(1), referring to the hay store, is set at zero during this calculation; it is then set at 1000.0 (line 59) as a convenient means of converting amounts of hay from kg to tonnes elsewhere in the program.
The number of stock classes is fixed by the specified flock structure (Fig. 2) and cannot be altered without making program changes. However, the initial numbers in each class can be varied between runs by expressing S(1) - S(19) as the required proportions and the total number of adult animals as C(81), and calculating actual numbers (line 64). The initial proportion in all adult classes, S(1), S(2) and S(3), is re-

Table 4 Parameters in the T array

Row	Column	Parameters
1		Times for general events as set out below
1	1	Calculation of herbage composition and animal nutrition
1	2	Changes in status of fodder unit
1	3	Changes in status of stock class
1	4	Allocation of mobs to fodder units
1	5	Allocation of stock classes to mobs and of mobs to fodder units
1	6	Assessment of objective function and changes in value of optimization parameter
1	7-19	Unused
2	1-19	Times for events concerning fodder units 1-19
3	1-19	Times for events concerning stock classes 1-19
4	1-19	Priority values for fodder units 1-19
5	1-19	Priority values for stock classes 1-19
6	1-19	Relative intakes, at last drafting, on fodder units 1-19

tained as RS (line 60), since this value determines the number of ewes to be culled and the number of weaners needed as replacements at the end of each year.

The next step is to allocate stock classes to mobs and mobs to fodder units, according to feed requirements and supplies. This necessitates a preliminary assessment, made on the basis of the simulation of a single day's grazing, in which all stock classes are put into a single mob (line 63), and distributed over all fodder units (line 55). The times in the first line of the T array are arranged to ensure that subroutine DRAFT (general event number 5) is called immediately afterwards to effect the redistributions. In all subsequent changes involving fodder units or stock classes, subroutine DRAFT is automatically called *before* the next day's grazing. Finally, the parameter C(22) is calculated (line 65) as the weight at birth of a single lamb from a ewe of normal mature weight; C(24) (line 66) is required for subsequent calculations of feed intake.

Start of each year

Each simulation run extends over NYR years (line 68). At the start of each year, the main program calls subroutine SEASON, which rewinds output files, writes output headings, and calculates the pattern of seasonal herbage growth. Subroutine OPTI is called at the start of the first year to calculate the initial value of the objective function; this is done by giving T(6,1) the smallest value in the first row of the T array.

During the run

The model proceeds through time from each general event to the next. The nature of the event is specified by the number N in location NT (20,1) (line 70) as determined previously in subroutine MEMO. The current time TT then becomes the time of this event (line 71). According to the value of N, control is transferred to the appropriate subroutine (line 72). Having carried out the required operations, the subroutine must call subroutine MEMO, specifying at what time in the future the same event is to occur again, before returning control to the main program. Subroutine MEMO accordingly rearranges the event times in the proper sequence, again placing the number of the next event in NT(20,1) ready for use by the main program. At the end of NYR years, following a call to subroutine OPTI, the program exits from the loop ending at statement 200 (line 86), and starts a new run at statement 2 (line 20) reading in the initial parameter values again. The optimization parameters are then changed (lines 43-44). The program terminates either when the specified number of runs, NRUN, has been reached, or when all possible combinations of optimization parameters have been examined.

Program EIEIO

```
 1          PROGRAM EIEIO
 2          COMMON/A/RD(6,20),DM(4,10),FM(10),SM(10),QM(10),TT,TYR,DFT
 3          COMMON/B/T(20,6),F(6,20),FL(5,3),Q(5),CFG(27,3),FG(3),TLU(10)
 4          COMMON/C/C(100),A(20),S(20),B(20),BN(20),AGE(20),W(20),
 5        * DI(20),EL(20),PJ(20),PK(20),VJ(20),VK(20),FI(20),TI(20),TK(20)
 6          COMMON/D/D(20),DK(20),E(20),BG(20),MJ(20),MK(20),NT(20,6)
 7          COMMON/Y/S0(2),S1(2),S2(2),B1(2),B2(2),B0(2),TMID(2)
 8          COMMON/O/OJ(20),OK(20),OW(20),OV(20),XOP(40),COP(20),DOP(20),
 9        * IOP(20),NOP,NRUN,NDO,IYR,IO,IRUN,IDO,IREV,IMP,OO,OOO,OEI,OEIT
10          COMMON/M/NFG(20),IJ(20),IK(20),KRE(20),KY(20)
11          COMMON/N/NJ,NK,NL,NM,NP,NDP,NYR,LU,LFG,L1,L2,L3,L4,L5,L6,L7,L8
12          COMMON/G/IIG,IG(125),LG(8),LO,LDOT
13          DATA LDOT,LO,LG/'.',' ','1','2','3','4','5','6','7','8'/
14          DATA IO,IRUN,IDO,IREV,IMP,OO,OOO /5*0,0.0,-999.0/
15    101   FORMAT(//20I4)
16    102   FORMAT(//20F4.0)
17          DO 1 I=1,160
18      1   OJ(I)=0.0
19          LU=11
20      2   READ(LU,101) LFG
21          READ(LU,101) L1,L2,L3,L4,L5,L6,L7,L8
22          READ(LU,102) TLU
```

```
      READ(LU,101)NRUN
      IRUN=IRUN+1
      IF(IRUN.GT.NRUN) STOP
      DO 3 I=1,160
    3 RD(I,1)=0.0
      DO 4 I=1,80
    4 D(I)=0.0
      DO 5 I=1,160
    5 MJ(I)=0
      READ(LU,101) NFG,IJ,IK,KRE,KY,NK,NL,NM,NP,NDP,NYR
      READ(LU,102) ((T(N,I),N=1,20),I=1,3),((F(L,J),J=1,20),L=1,6)
      READ(LU,102) FL,Q,C,A,S,B,BN,AGE,W,DI,EL
      READ(LU,102) PJ,PK,VJ,VK,FI,TI,TK
      READ(LU,102) S0,S1,S2,B1,B2,B0,TMID
      READ(LU,101) IOP,NOP,NDO
      IF(IRUN.GT.1) GO TO 7
      DO 6 I=1,NOP
    6 COP(I)=C(IOP(I))
      READ(LU,102) DOP
      GO TO 10
    7 DO 8 I=1,NOP
    8 C(IOP(I))=COP(I)
   10 DO 12 I=1,6
      T(20,I)=9999.0
   12 NT(20,I)=20
      DO 14 I=1,6
   14 CALL MEMO (I,1,T(I,1))
      NJ=C(8)
      A(1)=0.0
      RA=0.0
      DO 15 J=1,NJ
      CALL MEMO (J,2,T(J,2))
      MJ(J)=1
   15 RA=RA+A(J)
      DO 16 J=2,NJ
   16 A(J)=A(J)*A(20)/RA
      A(1)=1000.0
      RS=S(1)+S(2)+S(3)
      DO 20 K=1,NK
      CALL MEMO (K,3,T(K,3))
      MK(K)=1
   20 S(K)=S(K)*C(81)/RS
      C(22)=0.10*BN(20)
      C(24)=C(24)*BN(20)**0.75
      TT=0.0
      DO 200 IYR=1,NYR
      CALL SEASON
   30 N=NT(20,1)
      TT=T(N,1)
      GO TO(31,32,33,34,35,36) N
   31 CALL FODDER
      CALL STOCK
      GO TO 30
   32 CALL RATION
      GO TO 30
   33 CALL REPRO
      GO TO 30
   34 CALL DRAFT(0)
      GO TO 30
   35 CALL DRAFT (1)
      GO TO 30
   36 CALL OPTI
      IF(TT.LT.365.0) GO TO 30
  200 CONTINUE
      REWIND LU
      GO TO 2
      STOP
      END
```

4 Subroutine SEASON

This subroutine is called at the start of each year of simulation to prepare the output files for the tabulation of the current year's results and to prepare the pasture growth parameters for their use in subroutine FODDER. The set of these parameters appropriate for the season specified by the user is read in from the input file and transferred to arrays in a form which can be used for calculating daily growth rates of herbage. If the user wishes to incorporate an alternative sub-model for predicting pasture growth, this part of the subroutine (lines 51-66) may be omitted. The variables in the subroutine are defined in Table 5.

Lines 36-43 rewind the output files at the start of each optimization run. These steps restrict the otherwise large volume of output to the final run, but should be deleted if the full output is required. Lines 44-50 provide explanatory headings on each of the output files.

The subroutine then reads the data from the LFG file into the CFG array specifying the characteristics of pasture growth for the particular sward and pattern of seasonal conditions in the sequence set out in the NFG array (lines 54-56). For instance, if NFG(1) is set at 4, the program at the start of year 1 reads in the fourth pattern, overwriting the first three.

The CFG array consists of 3 rows of data, each containing a value for the start of the calendar year and at 14-day intervals thereafter.

Row 1, CFG(NP,1): Ceiling yield, or the maximum quantity of herbage dry matter (kg ha^{-1}) which can be sustained in period NP. This parameter is assumed to be a function of sward type and solar radiation.

Row 2, CFG(NP,2): The relative growth rate at the point of maximum absolute growth rate or when the quantity of herbage is half the ceiling yield. This parameter is assumed to be a function of sward type, temperature and soil moisture.

Row 3, CFG(NP,3): Relative maturation rate, governing the proportion of material in each digestibility class which moves daily to the next lower class. In subroutine FODDER, this

Table 5 Definition of variables in subroutine SEASON

Variable name	Definition
CFG(NP,N)	Fodder growth factor N for period NP
FG(N)	Current value of fodder growth factor N
IFG	Current value of NFG(IYR)
IP	Number of the initial period for the current year
IRUN	Number of current optimization run
IYR	Current year of optimization run
LFG	Logical unit for input of seasonal growth factors
NDP	Number of days in each period NP
NFG(IYR)	Pattern of seasonal growth factors selected for year IYR

seasonal factor is multiplied by a class factor to calculate the maturation rate for each herbage class.
These values are converted to daily increments (lines 61-63), in which form they are used by subroutine FODDER to generate daily pasture growth. On the first day of the year, the initial values are used; on subsequent days, the increments for the appropriate period are added to the previous day's figures.
The growth pattern can start on any day of the calendar year, as specified by C(1). An example of a predicted seasonal pattern of herbage yield in the absence of grazing animals is given in Section 12.
It is convenient for the times shown in the T-array to refer to the time from the start of the current 12-month period, rather than to increase indefinitely. Hence, for every year after the first, those events which are still operative (rows 1-3, with T < 999.0) are decreased by 365 days (line 70). The convention is adopted that events which are scheduled to occur at later than 999 days will in fact never happen.

Program subroutine SEASON

```
      SUBROUTINE SEASON
      COMMON/A/RD(6,20),DM(4,10),FM(10),SM(10),QM(10),TT,TYR,DFT
      COMMON/B/T(20,6),F(6,20),FL(5,3),Q(5),CFG(27,3),FG(3),TLU(10)
      COMMON/C/C(100),A(20),S(20),B(20),BN(20),AGE(20),W(20),
     * DI(20),EL(20),PJ(20),PK(20),VJ(20),VK(20),FI(20),TI(20),TK(20)
      COMMON/D/D(20),DK(20),E(20),BG(20),MJ(20),MK(20),NT(20,6)
      COMMON/Y/S0(2),S1(2),S2(2),B1(2),B2(2),B0(2),TMID(2)
      COMMON/O/OJ(20),OK(20),OW(20),OV(20),XOP(40),COP(20),DOP(20),
     * IOP(20),NOP,NRUN,NDO,IYR,IO,IRUN,IDO,IREV,IMP,OO,OOO,OEI,OEIT
      COMMON/M/NFG(20),IJ(20),IK(20),KRE(20),KY(20)
      COMMON/N/NJ,NK,NL,NM,NP,NDP,NYR,LU,LFG,L1,L2,L3,L4,L5,L6,L7,L8
  103 FORMAT(/-2P,26F3.0,F2.0/3P,26F3.0,F2.0/1P,26F3.0,F2.0)
  110 FORMAT(//' RUN',I4,3X,'YEAR',I4,3X,'FOD.GROWTH',I4/)
  112 FORMAT(' FODDER ON SELECTED FODDER UNIT, J=2'/
     * ' DAY',7X,'FEED DM/HA (EACH DIG.CLASS)',10X,'RELATIVE INTAKE
     * (EACH DIG.CLASS)',4X,'YIELD  GROWTH  DECAY
     * 6X,6I6,2X,6I6,4X,'FG(1)   FG(2)   FG(3)     YIELD  GROWTH'/)
  113 FORMAT(' NUTRITION OF SELECTED STOCK CLASS, K=1'//
     * ' DAY PRIORITY POTL.  ACTUAL   HAY   INTAKE DIGEST- METAB.
     * MILK  ENERGY WEIGHT  BODY'8X,' DDM INTAKES FOR
     * EACH CLASS'/14X,'INTAKE INTAKE INTAKE   DDM IBILITY ENERGY
     * ENERGY SURPLUS GAIN  WEIGHT
  114 FORMAT(' QUANTITY OF HERBAGE IN EACH DIGESTIBILITY CLASS',
     * '(CUMULATIVE,ALL PADDOCKS,KG/HA)'/4X,'0',16X,'2000',16X,'4000',
     * 15X,'6000',16X,'8000',16X,'10000 KG/HA'/)
  115 FORMAT(' BODYWEIGHT OF EACH STOCK CLASS'//4X,'0',18X,
     * '10',18X,'20',18X,'30',18X,'40',18X,'50 KG'/)
  116 FORMAT(' RATION'/' DAY UNIT STATUS',15X,'FODDER IN EACH CLASS'/
     * 20X,8I7/)
  117 FORMAT(' REPRO'/' DAY CLASS STATUS DAY AGE',20X,'SHEEP NUMBER'
     * ,40X,'BODYWEIGHT'/26X,8I6,6X,8I6/)
  118 FORMAT(' DRAFT'/' DAY UNIT   STATUS',8X,'FEED (F(6,J)) AND MOB
     * NO.(MJ) ON EACH UNIT',8X,'PRIORITY (PKK) AND MOB NO.(MK) OF
     *  EACH CLASS'//7X,6I2,6I8,2X,8I7/)
      IF (IYR.GT.1) GO TO 11
      REWIND L1
      REWIND L2
      REWIND L3
      REWIND L4
      REWIND L5
      REWIND L6
      REWIND L7
      REWIND L8
   11 WRITE(L2,112) (L,L=1,6),(L,L=1,6)
      WRITE(L3,113) (L,L=1,8)
      WRITE(L4,114)
      WRITE(L5,115)
      WRITE(L6,116) (L,L=1,6)
      WRITE(L7,117) (L,L=1,8),(L,L=1,8)
      WRITE(L8,118) (L,L=1,6),(L,L=1,6),(L,L=1,8)
      REWIND LFG
      IFG=NFG(IYR)
      DO 21 I=1,IFG
   21 READ (LFG,103) CFG
      TP=C(1)/NDP+1.0
      IP=TP
      DO 23 II=1,3
      FG(II)=(IP+1.0-TP)*CFG(IP,II)+(TP-IP)*CFG(IP+1,II)
      DO 22 IL=1,NP
      I=NP+1-IL
      CFG(I+1,II)=(CFG(I+1,II)-CFG(I,II))/NDP
   22 CONTINUE
      CFG(1,II)=0.0
   23 FG(II)=FG(II)-CFG(IP+1,II)
      WRITE(L1,110) IRUN,IYR,IFG
      IF(IYR.EQ.1) RETURN
      DO 25 I=1,60
      IF(T(I,1).LT.999.0) T(I,1)=T(I,1)-365.0
   25 CONTINUE
      RETURN
      END
```

5 Subroutine MEMO

Both the scheduling of events and the ranking order of the various priorities are arranged in the correct sequence by subroutine MEMO, using the same algorithm. Since both operations are frequently required at the same time, it is convenient to handle them within the one subroutine. The sets of parameters dealt with are shown in Table 4, with each set forming one row (second subscript) of the T array. The sequences start at the lowest value and progress to the highest; where the sequence is required to start with the highest value and continue in descending numerical order, as in rows 5 and 6, a minus sign is attached to these parameters, so that the largest value becomes the most highly negative. Variables in the subroutine are defined in Table 6.

Each call made to subroutine MEMO from some other part of the program either re-sets the time for an event or changes a priority value, and in either case three parameters must be specified in line 1 (Fig. 5):

a. the column number I of the T array, (I = 1,19),
b. the row number II of the T array, (II = 1,6), and
c. the new time for the event, or the new priority value, TNI.

Since these formal parameters, I, II and TNI, would transmit incorrect values back to the calling routine if they were altered during the course of this subroutine, the equivalent parameters, N, NN and TN respectively, must be used in their place (lines 12-14).

Subroutine MEMO does not sort the T array; instead, it arranges numbers in a corresponding array, NT, which has the same dimensions and refers to the same parameters. Each row of the NT array may be looked upon as a linked set of pointers, with the last column (column 20) taken as a convenient reference point. The procedure for setting up and altering the NT array is illustrated in Fig. 6. The T array is read in during the initialization stage of the main program, with a very large value set in column 20, while all columns except the last in each row of the NT array are set at zero.

A typical row, NN, of the T and NT arrays might then look like this:

Table 6 Definition of variables in subroutine MEMO

Variable name	Definition
IJ(J)	Status of fodder unit J
IK(K)	Status of stock class K
N1	
N2	Pointers to column numbers in sorting process
NN	Column no. in T and NT arrays
NT(N,NN)	Array of pointers to corresponding parameters in T array
PJ(I)	Priority value of fodder unit with status I = IJ(J)
PK(I)	Priority value of stock class with status I = IK(K)
T(N,NN)	Array of events, priorities and relative intakes
TN	New time for an event or priority value

N	1	2	3	4	-----	20
T(N,NN)	2.2	1.6	7.5	1.9	-----	9999.9
NT(N,NN)	0	0	0	0	-----	20

The first call involving row NN = II comes from the main program, with N = I = 1 and TN = 2.2. Since NT(N,NN) = 0, control is transferred at line 15 to line 21. A search is then carried out (lines 22-24) to find numbers N1 and N2 such that T(N1,NN) < T(N,NN) < T(N2,NN), starting at column 20:

N2 = 20
N1 = N2 = 20
N2 = NT(20,NN) = 20
T(20,NN) < 2.2? No.

Hence the following changes are made (lines 25-27):

NT(20,NN) = N = 1
NT(1,NN) = N2 = 20
T(1,NN) = 2.2

with the result:

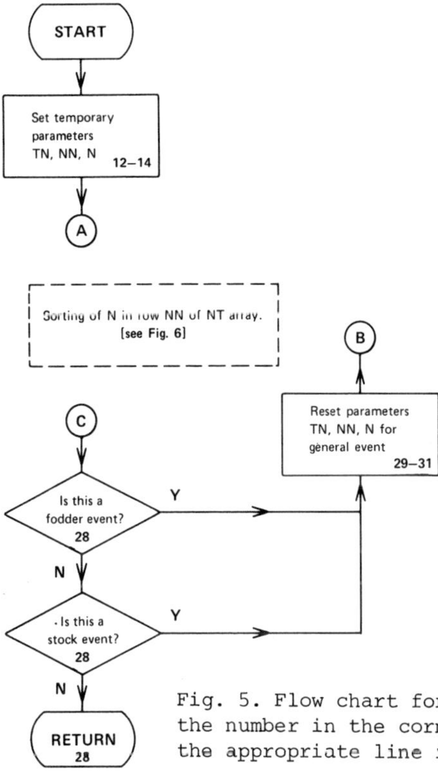

Fig. 5. Flow chart for subroutine MEMO (part 1); the number in the corner of each box indicates the appropriate line in the listing.

N	1	2	3	4	-----	20
T(N,NN)	2.2	1.6	7.5	1.9	-----	9999.9
NT(N,NN)	20	0	0	0	-----	1

Now NT(20,NN) = 1, i.e. it points to column 1, and NT(1,NN) = 20, i.e. it points to column 20, these being the only members of the set which have been sorted so far. Expressed diagrammatically in the order of the searching process:

N	20	1
NT(N,NN)	1	20

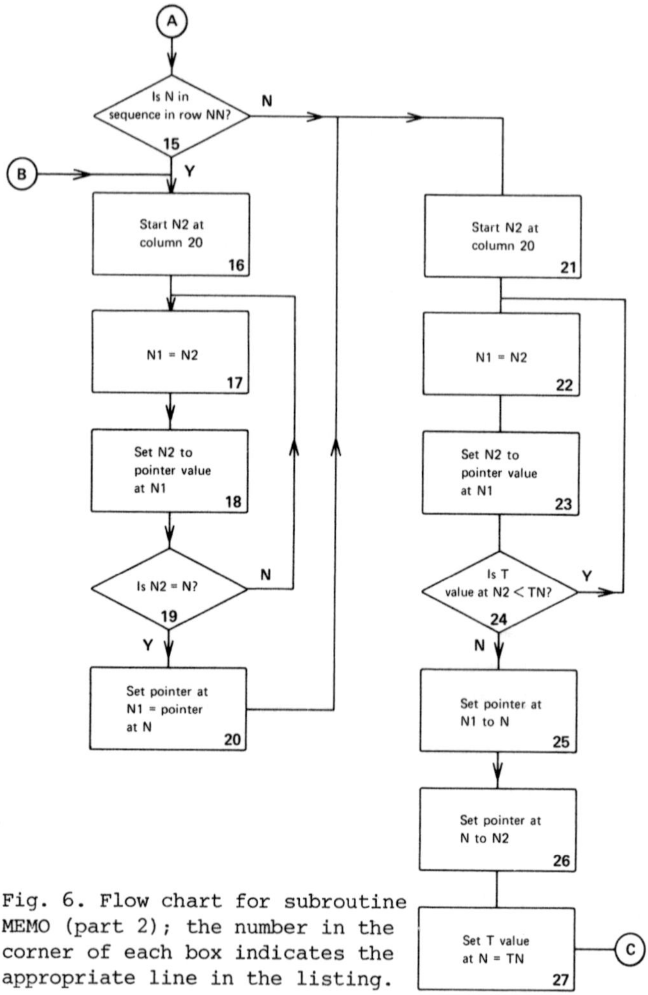

Fig. 6. Flow chart for subroutine MEMO (part 2); the number in the corner of each box indicates the appropriate line in the listing.

Column 20 therefore represents the end of the linked chain, as well as the beginning. The T(N,NN) value is not changed at this stage.
At the next call to MEMO, N = I = 2 and TN = 1.6; the search, again starting at line 21, runs as follows:

```
N2 = 20
N1 = N2 = 20
N2 = NT(20,NN) = 1
T(1,NN) = 2.2
2.2 < 1.6? No
```

Hence

```
NT(20,NN) = 2
NT(2,NN)  = 1
```

and the chain becomes

```
N            20    2    1

NT(N,NN)      2    1   20
```

With N = 3 and TN = 7.5, it is established in the same way that T(1,NN) = 2.2 < TN, but T(20,NN) > TN, so that

```
N            20    2    1    3

NT(N,NN)      2    1    3   20
```

Finally, T(4,NN) is placed between T(2,NN) and T(1,NN):

```
N            20    2    4    1    3

NT(N,NN)      2    4    1    3   20
```

A complete set of linkages has now been formed, with each relevant member of the row having one, and only one, pointer directed towards it. As a result, the events in the T array are now linked in ascending order as follows:

N	1	2	3	4	-----	20
T(N,NN)	2.2	1.6	7.5	1.9	-----	9999.9
NT(N,NN)	3	4	20	1	-----	2

After the NT array has been set up in this way, the procedure for changing the ranking order is slightly more complex, since each member of the array now has 2 pointers associated with it, i.e. a pointer towards it from the next lower member of the series, as well as its own pointer directed towards the next higher member. Suppose in the present example that a call to

MEMO re-schedules the time of the next event of T(2,NN) from 1.6 to, say, 2.0. Using the method just described, the successive calculations would be:

N2 = 20
N1 = N2 = 20
N2 = NT(20,NN) = 2
T(2,NN) = 1.6
1.6 < 2.0? Yes
N1 = 2
N2 = NT(2,NN) = 4
T(4,NN) = 1.9
1.9 < 2.0? Yes
N1 = 4
N2 = NT(4,NN) = 1
T(1,NN) = 2.2
2.2 < 2.0? No

Hence

NT(4,NN) = 2
NT(2,NN) = 1
T(2,NN) = 2.0

with the result:

N	1	2	3	4	-----	20
T(N,NN)	2.2	2.0	7.5	1.9	-----	9999.9
NT(N,NN)	3	1	20	2	-----	2

But now there are 2 pointers towards column 2, and none towards column 4, and this must be rectified. Since T(4,NN) now has the lowest value of the set, NT(20,NN) should be set at 4; that is, the pointer directed towards the column being changed should be diverted to the next higher member of the series, which is the column to which the pointer at the column being changed is directed. This is carried out by making a preliminary search of the NT array (lines 16-19), since NT(N,NN) is no longer zero:

N2 = 20
N1 = N2 = 20
N2 = NT(20,NN) = 2
N2 = N? Yes

Accordingly (line 20):

NT(20,NN) = NT(2,NN) = 4

The set of linkages is now as follows:

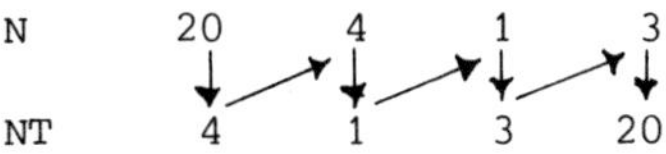

Although the pointer at column 2 is still directed towards column 4, it is now ineffective, because column 2 cannot be reached from any other point in the chain. Continuing, a search is made as before, with N = 2 and TN = 2.0, (lines 21-24) to give:

N2 = 20
N1 = N2 = 20
N2 = NT(20,NN) = 4

T(4,NN) = 1.9
1.9 < 2.0? Yes
N1 = 4
N2 = NT(4,NN) = 1

T(1,NN) = 2.2
2.2 < 2.0? No
NT(4,NN) = 2
NT(2,NN) = 1

The linkages are now as follows:

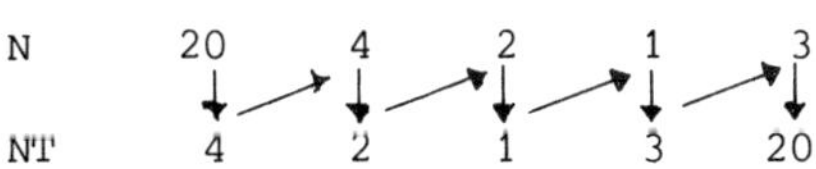

The change to the NT array is now complete:

N	1	2	3	4	-----	20
T(N,NN)	2.2	2.0	7.5	1.9	-----	9999.9
NT(N,NN)	3	1	20	2	-----	4

For this simple case, the steps involved may appear unnecessarily complicated, but they nevertheless represent a systematic and efficient means of dealing automatically with all the eventualities which can arise.

Each time a fodder unit or stock class event is rescheduled, the time for the subroutine to be called again must also be updated as shown in Fig. 6. If a fodder unit is involved (N = 2), the number of the unit to be dealt with next is specified as NT(20,2), and the time of this event is T(NT(20,2),2) (line 31). This value is therefore given to the time for the next fodder event, T(2,1); and the first row is then sorted accordingly, by repeating the cycle with the new values for N, NN and TN (lines 29-31).

When a fodder unit or stock class changes in status, its priority value often changes as well, necessitating a call to subroutine DRAFT scheduled from the appropriate routine. Subroutine DRAFT requires rows 4-6 of the T array to be arranged in order, which it accomplishes by a series of calls to subroutine MEMO.

Program subroutine MEMO

```
 1          SUBROUTINE MEMO (I,II,TNI)
 2          COMMON/A/RD(6,20),DM(4,10),FM(10),SM(10),QM(10),TT,TYR,DFT
 3          COMMON/B/T(20,6),F(6,20),FL(5,3),Q(5),CFG(27,3),FG(3),TLU(10)
 4          COMMON/C/C(100),A(20),S(20),B(20),BN(20),AGE(20),W(20),
 5        * DI(20),EL(20),PJ(20),PK(20),VJ(20),VK(20),FI(20),TI(20),TK(20)
 6          COMMON/D/D(20),DK(20),E(20),BG(20),MJ(20),MK(20),NT(20,6)
 7          COMMON/Y/S0(2),S1(2),S2(2),B1(2),B2(2),B0(2),TMID(2)
 8          COMMON/O/OJ(20),OK(20),OW(20),OV(20),XOP(40),COP(20),DOP(20),
 9        * IOP(20),NOP,NRUN,NDO,IYR,IO,IRUN,IDO,IREV,IMP,OO,OOO,OEI,OEIT
10          COMMON/M/NFG(20),IJ(20),IK(20),KRE(20),KY(20)
11          COMMON/N/NJ,NK,NL,NM,NP,NDP,NYR,LU,LFG,L1,L2,L3,L4,L5,L6,L7,L8
12          TN=TNI
13          NN=II
14          N=I
15    1     IF(NT(N,NN).EQ.0) GO TO 4
16    2     N2=20
17    3     N1=N2
18          N2=NT(N1,NN)
19          IF (N2.NE.N) GO TO 3
20          NT(N1,NN)=NT(N,NN)
21    4     N2=20
22    5     N1=N2
23          N2=NT(N1,NN)
24          IF(T(N2,NN).LE.TN) GO TO 5
25          NT(N1,NN)=N
26          NT(N,NN)=N2
27          T(N,NN)=TN
28          IF(NN.LT.2.OR.NN.GT.3) RETURN
29          N=II
30          NN=1
31          TN=T(NT(20,N),N)
32          GO TO 2
33          END
```

6 Subroutine FODDER

This subroutine calculates the following three sets of variables daily:
a. the growth of herbage in each paddock and the distribution of this between five herbage classes which range in digestibility from 0.8 to 0.4, and the weight of herbage in each class after correcting for growth, maturation and decay,
b. the relative intake achievable from each paddock as a function of the availability and digestibility of herbage in each class, and
c. the accumulated relative intakes and diet digestibilities for all paddocks grazed by each mob, for the calculation of the weighted mean values of these variables.
The variables used in the subroutine are defined in Table 7.
Each day, before herbage growth is calculated, the values of ceiling yield, FG(1), relative growth rate, FG(2), and senescence rate, FG(3), are incremented at line 16 by the appropriate daily changes in the CFG array, as calculated in SEASON.
Between lines 17 and 23 several variables that will be recalculated daily are given zero values.
The hay store (fodder unit 1) is not considered in this subroutine. For each of the other units, the status (I) and mob (M) are identified.
Daily herbage growth is generated (line 28) from a logistic function of ceiling yield, relative growth rate, and the weight of green herbage, FO. It is assumed that FO does not fall below a minimum value, C(11), which represents plant reserves inaccessible to the grazing animal.
The growth function (Fig. 7) was derived as follows:

In general:

$$g = kh(c - h)$$

At the point of maximum growth rate:

$$g = \frac{kc}{2}\left(c - \frac{c}{2}\right) = \frac{kc^2}{4}$$

Table 7 Definition of variables in subroutine FODDER

Variable name	Definition
A(J)	Area of fodder unit J (ha)
CFG(NP,N)	Growth and maturation factors for NP periods of the year; used to calculate daily values of FG(N)
DM(1,M)	Sum of (relative intake x area) for all fodder units grazed by mob M
DM(2,M)	Sum of (relative intake^2 x area) for all units grazed by mob M
DM(3,M)	See STOCK
DM(4,M)	See DRAFT
F1	Herbage leaving a class by maturation (kg DM ha^{-1} day^{-1})
F2	Herbage entering a class by maturation (kg DM ha^{-1} day^{-1})
F(L,J)	Weight of fodder in class L (kg DM ha^{-1})
F(6,J)	Total fodder in unit J (kg DM ha^{-1})
F(6,20)	Fodder on whole farm (kg DM ha^{-1})
FF	Daily herbage growth (kg DM ha^{-1})
FG(1)	Ceiling yield of herbage (kg DM ha^{-1})
FG(2)	Relative growth rate at the point of maximum growth rate
FG(3)	Seasonal factor modifying the rate of maturation of herbage
FL(L,1)	Proportion of day's growth that enters class L
FL(L,2)	Class factor for calculating daily maturation in class L
FL(L,3)	1.0 - (proportion of each class disappearing daily by decay)
FO	Weight of green herbage on which growth rate depends (kg DM ha^{-1})
IJ(J)	Status of fodder unit J with respect to grazing or conservation
IP	1 + the number of the current growth period
MJ(J)	Reference number of the mob that is currently grazing fodder unit J
NDP	Number of days per growth period
NJ	Number of fodder units
NL	Number of digestibility classes of herbage
NP	Number of growth periods
Q(L)	Digestibility coefficient of herbage in class L
QD	Contribution of each paddock to the mean digestibility of the diet eaten by mob M
QF	Digestibility of herbage shut up for fodder

Table 7 (continued)

Variable name	Definition
	conservation
QM(M)	Mean digestibility of the diet eaten by mob M
RD(L,J)	Proportion of potential dry matter intake achievable from class L in unit J
RD(6,J)	Proportion of potential dry matter intake achievable from paddock J
RLL	Cumulative availability factor for all higher herbage classes
SM(M)	No. of sheep in mob M
TT	No. of days from start of run
TYR	Time of year (days from January 1)

and

$$r = \frac{2g}{2} = \frac{kc}{2}$$

therefore

$$k = \frac{2r}{c}$$

therefore at all points

$$g = \frac{2rh}{c}(c - h) = 2rh(1 - \frac{h}{c})$$

where

g = daily growth (kg DM ha^{-1})
h = green herbage present (kg DM ha^{-1})
c = ceiling yield of herbage (kg DM ha^{-1})
r = relative growth rate at the point of maximum growth rate (kg kg^{-1})
k is a constant.

The daily recalculation of the weight of herbage in digestibility class L (line 33) takes the following changes into account.
a. New growth enters class L (for L = 2 - 5). This amount is a proportion, FL(L,1), of the total daily growth, FF.
b. Green material moves to the next lower digestibility class (L-1) by maturation for L = 2 - 5. This amount, F1, is calculated in line 32 as a proportion, FL(L,2), of the weight in class L, modified by a seasonal factor FG(3) which increases

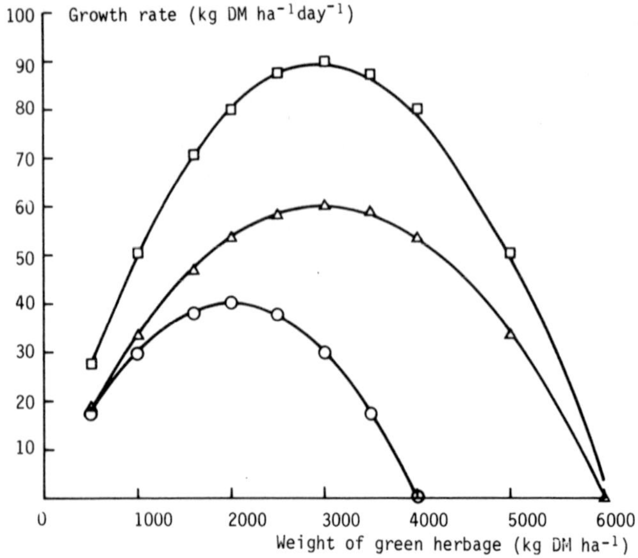

Fig. 7. Growth rate of herbage at different levels of herbage weight, for specified values of relative growth rate (r) and ceiling yield (c). ○——○, r=0.02, c=4000; △——△, r=0.02, c=6000; □——□ r=0.03, c=6000.

rapidly at the onset of senescence.
c. Material enters class L by maturation from the class above (except when L = 5). This amount left class (L+1) as F1 but enters class L as F2. The net gain by maturation is (F2-F1).
d. Material is lost from the class by decay. This is calculated as a proportion (1-FL(L,3)) of the weight of herbage in the class.

The weights of herbage in all classes are summed for each fodder unit as F(6,J) in line 35 and from these totals the mean weight over the whole farm is calculated as F(6,20) in line 36.
By choosing suitable parameters for the CFG array, any required pattern of herbage growth may be produced. We have constructed a number of arrays to simulate a range of seasons typical of this environment. The predicted yields of herbage (see Section 12) agree well with those measured in the field. For other environments, of course, the appropriate changes must be made to the CFG array.
Lines 40-48 calculate the relative intake RD(6,J) for each paddock. This is the proportion of an animal's potential intake

that is achievable from the paddock and is used in subroutine STOCK to calculate the actual daily intake. It is assumed that the potential intake is progressively satisfied from successive classes of herbage in descending order of digestibility.
The contribution, RD(L,J), of class L to the relative intake depends on its digestibility, the amount present and the proportion of the potential intake left unsatisfied by higher digestibility classes. The experimental results of Willoughby (1959) and McKinney et al. (1970) indicate a general relationship between intake and herbage availability which supports the assumption that for herbage with a digestibility of 0.8,

$$\frac{df}{dh} = a(1 - f)$$

i.e. $f = 1 - e^{-ah}$

where

f = the relative intake
h = weight of herbage (kg ha^{-1})
a is a constant.

However a value of f calculated from a function of this general form will apply only over a time interval sufficiently short for the intake of herbage to have no significant effect on the value of h. We have found that the time interval of one day used in this model between calculations is adequate under extensive grazing but is too long for this condition to hold under intensive rotational grazing systems. Rather than increase the frequency of calculations we have made the further assumption that the rate of change of intake with respect to availability is itself affected by the availability of herbage,

i.e. $$\frac{df}{dh} = ah(1 - f)$$

Predictions based on a function of this form are in better agreement with experimental results but accurate information is sparse, particularly at low levels of availability.
For a particular herbage class i with digestibility q_i,

$$\frac{df_i}{dh_i} = ah_i(1 - \frac{f_i}{c_i})$$

where

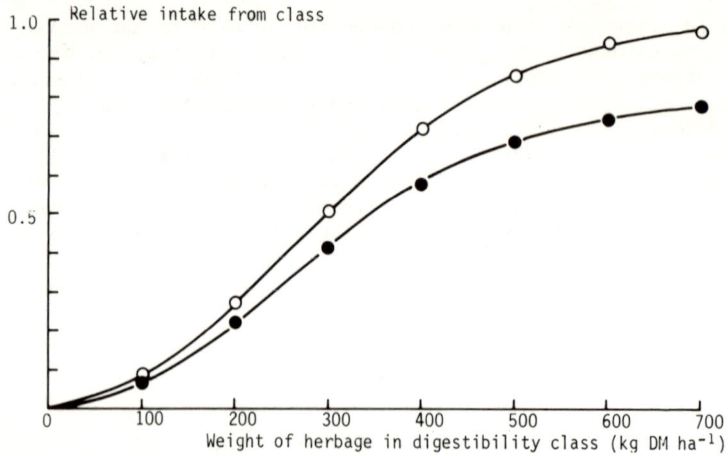

Fig. 8. Relative intake from an individual class of herbage in relation to the weight of material in the class for specified values of digestibility (q) and with a=1.6 x 10^{-5}. ○——○, q=0.8; ●——●, q=0.7.

$$c_i = 1 - 2(0.8 - q_i)$$

$$\therefore f_i = c_i(1 - e^{-ah_i^2/2c_i})$$

The form of this function is illustrated in Fig. 8.

The total relative intake is calculated as the summation of f_i, where f_i is the contribution from each herbage class after correcting for the proportion of relative intake already satisfied by higher digestibility classes.

$$\sum_{i=1}^{n} f_i = \sum_{i=1}^{n} \{(1 - \sum_{i=1}^{n} f_{i-1}) c_i (1 - e^{-ah_i^2/2c_i})\}$$

A numerical example of these calculations is shown in Table 8.

It is apparent from Fig. 8 that even at low levels of availability, the predicted relative intake of less digestible material will be lower than that of more digestible material. This differs from the result to be expected from penned animals offered small amounts of food, but is consistent with the observations (a) that intake by grazing animals on sparse pasture is limited by the time available for grazing (Arnold 1964)

Table 8 Numerical example of the contribution of each herbage class to the relative intake achievable from a paddock (using $a=1.6 \times 10^{-5}$)

Variable	Name	Herbage class (L)				
		5	4	3	2	1
Digestibility (q_i)	Q(L)	0.8	0.7	0.6	0.5	0.4
Weight (h_i) of herbage (kg DM ha^{-1})	F(L,J)	200	200	300	400	600
$c_i = 1 - 2(0.8 - q_i)$	RX	1.0	0.8	0.6	0.4	0.2
$f_i = c_i(1 - e^{-ah_i^2/2c_i})$		0.274	0.264	0.419	0.384	0.200
$(1 - \sum_{i=1}^{n} f_{i-1})\ f_i$	RD(L,J)	0.274	0.192	0.224	0.119	0.038
Total relative intake	RD(6,J)					0.847

and (b) that the less digestible the diet, the greater the proportion of this time that will be spent on chewing the food, either during eating or ruminating (Balch 1971).
Since the sheep classes graze as mobs, each of which may have access to several fodder units, it is necessary in subroutine STOCK to calculate the mean relative intake for each mob. This mean is weighted for the respective areas of the individual fodder units and is represented by DM(2,M)/DM(1,M). The components of this ratio are calculated in lines 55 and 56. The mean digestibility of the diet eaten by each mob, QM(M), is calculated in a similar way. The contribution of each herbage class in each paddock to the relative intake of digestible material is accumulated as QD at line 49. These values are summed for all paddocks grazed by each mob at line 57 and the mean digestibility, weighted for relative intake and area of paddock, is calculated at line 60.

Units shut up for hay (I = 3) are examined daily (line 39), and if the digestibility has fallen below a critical level, C(16), subroutine MEMO is called to schedule haymaking immediately.

Program subroutine FODDER

```
      SUBROUTINE FODDER
      COMMON/A/RD(6,20),DM(4,10),FM(10),SM(10),QM(10),TT,TYR,DFT
      COMMON/B/T(20,6),F(6,20),FL(5,3),Q(5),CFG(27,3),FG(3),TLU(10)
      COMMON/C/C(100),A(20),S(20),B(20),BN(20),AGE(20),W(20),
     * DI(20),EL(20),PJ(20),PK(20),VJ(20),VK(20),FI(20),TI(20),TK(20)
      COMMON/D/D(20),DK(20),E(20),BG(20),MJ(20),MK(20),NT(20,6)
      COMMON/Y/SO(2),S1(2),S2(2),B1(2),B2(2),BO(2),TMID(2)
      COMMON/O/OJ(20),OK(20),OW(20),OV(20),XOP(40),COP(20),DOP(20),
     * IOP(20),NOP,NRUN,NDO,IYR,IO,IRUN,IDO,IREV,IMP,OO,OOO,OEI,OEIT
      COMMON/M/NFG(20),IJ(20),IK(20),KRE(20),KY(20)
      COMMON/N/NJ,NK,NL,NM,NP,NDP,NYR,LU,LFG,L1,L2,L3,L4,L5,L6,L7,L8
  101 FORMAT(F5.0,2X,6F6.0,2X,6F6.2,2X,F7.0,2F7.3,2X,F7.0,F7.2)
      TYR=AMOD(TT-1.0+C(1),365.0)+1.0
      IP=(TYR-2.0)/NDP+2
      DO 2 N=1,3
    2 FG(N)=FG(N)+CFG(IP,N)
      DO 5 M=1,NM
      SM(M)=0.01
      QM(M)=0.0
      DO 5 N=1,3
    5 DM(N,M)=0.001
      DO 10 L=1,6
   10 F(L,20)=0.0
      DO 30 J=2,NJ
      QF=0.0
      F2=0.0
      FO=AMAX1(F(6,J)-F(1,J),C(11))
      FF=2.0*FG(2)*FO*DIM(1.0,FO/FG(1))
      F(6,J)=0.0
      DO 15 LL=1,NL
      L=NL+1-LL
      F1=FG(3)*FL(L,2)*F(L,J)
      F(L,J)=FL(L,3)*(F(L,J)-F1+F2+FL(L,1)*FF)
      F2=F1
      F(6,J)=F(6,J)+F(L,J)
      F(L+1,20)=F(L+1,20)+F(L,J)*A(J)/A(20)
      QF=QF+F(L,J)*Q(L)
   15 CONTINUE
      IF(IJ(J).EQ.3.AND.QF/F(6,J).LT.C(16)) CALL MEMO (J,2,TT)
      RLL=1.0
      QD=0.0
      RD(6,J)=0.0
      DO 20 LL=1,NL
      L=NL+1-LL
      RQ=1.0-2.0*(0.8-Q(L))
      RD(L,J)=RLL*RQ*(1.0-EXP(-0.5*(C(12)*F(L,J))**2.0/RQ))
      RLL=RLL-RD(L,J)
      RD(6,J)=RD(6,J)+RD(L,J)
      QD=QD+RD(L,J)*Q(L)
   20 CONTINUE
      IF(J.EQ.2.AND.AMOD(TT,TLU(2)).LT.0.9) WRITE(L2,101)TT,(F(L,J),
     * L=1,6),(RD(L,J),L=1,6),(FG(L),L=1,3),FO,FF
      M=MJ(J)
      IF(M.EQ.0) GO TO 30
      DM(1,M)=DM(1,M)+RD(6,J)*A(J)
      DM(2,M)=DM(2,M)+RD(6,J)*RD(6,J)*A(J)
      QM(M)=QM(M)+QD*A(J)
   30 CONTINUE
      DO 40 M=1,NM
   40 QM(M)=QM(M)/DM(1,M)
      RETURN
      END
```

7 Subroutine STOCK

Subroutine STOCK calculates the intake of metabolizable energy (ME) by each class of animals, subtracts requirements for maintenance, foetal growth and lactation, and converts the remainder to body weight gain. The weight of herbage in each fodder unit is then corrected for the amount eaten and trampled during grazing. The variable names used in the subroutine are defined in Table 9.

The upper limit to the voluntary intake of food is assumed to depend on an animal's stage of development rather than on age or weight alone. This assumption is consistent with the observations of Allden (1968) on young sheep suffering and recovering from nutritional deprivation and with the results of Donnelly et al. (1974) from mature sheep. The index of development, termed normal body weight, BN(K), is calculated daily for each class (line 25) as the minimum of two parameters: (i) the highest weight attained so far by the animals in that class and (ii) the weight of a well-grown animal of the same age, BNMAX, calculated at line 24 from the following function (illustrated in Fig. 9) which is based on Brody (1945):

$$w = a - (a-b)e^{-kt}$$

where

w = weight of well-grown animal (kg)
a = mature weight (upper limit of normal body weight) (kg)
b = birth weight, assumed to be 0.1a (kg)
k = growth constant
t = age (days)

Normal body weight, as a proportion of its upper limit (called mature body weight), is used at line 26 to calculate the potential intake, DK(K), which is defined as the amount of dry matter that will be eaten per head if intake is not limited by the availability or digestibility of food. The following function (illustrated in Fig. 10) is similar in shape to the one suggested by Blaxter (1968), but we have found that a quadratic relationship fits our own data rather better

Table 9 Definition of variables in subroutine STOCK

Variable name	Definition
A(J)	Area of fodder unit J (ha)
AGE(K)	Age of stock in class K (days)
AGEW	Age factor for calculating wool growth
B(K)	Mean body weight of stock in class K (kg)
B1(K)	Weight of a single foetus in a ewe in class K (kg)
B2(K)	Weight of a twin foetus in a ewe in class K (kg)
BC	Critical body weight in relation to death from undernutrition (kg)
BG(K)	Daily weight gain (kg)
BG1	Daily weight gain of a single foetus (kg)
BG2	Daily weight gain of a twin foetus (kg)
BGY	Daily weight gain of a single lamb from a ewe of normal mature weight (kg)
BN(K)	Normal body weight of stock in class K (kg)
BN(20)	Normal weight of a mature animal (limit of BN(K)) (kg)
BNMAX	Weight of a well-grown animal as a function of its age (kg)
BO(K)	Function of metabolic size and condition of ewe at lambing (kg)
D(K)	Intake of digestible dry matter (DDM) per animal in class K (kg)
DF	Intake of dry matter from herbage per animal (kg)
DFT	Total dry matter intake from herbage for all animals (kg)
DHAY	Dry matter intake of hay per animal (kg)
DHAYT	Total weight of hay dry matter consumed (kg)
DI(K)	Factor for changing potential intake during lactation or suckling
DJJ	Potential intake of fodder dry matter from paddock J (kg ha^{-1})
DK(K)	Potential intake of dry matter by stock in class K
DM(1,M)	Sum of (relative intake x area) for all paddocks grazed by mob M
DM(2,M)	Sum of (square of relative intake x area) for all paddocks grazed by mob M
DM(3,M)	Sum of potential intakes for all animals in mob M
E(K)	Intake of metabolizable energy per animal (MJ day^{-1})
EB	Net energy required for basal metabolism (MJ day^{-1})

Table 9 (continued)

Variable name	Definition
EL(K)	ME content of milk produced per ewe (negative) or consumed per lamb (positive) (MJ day^{-1})
ELB	Proportion by which potential milk yield is reduced to actual yield
ELPOT	Potential milk yield per ewe (MJ day^{-1})
ELMAX	Actual milk produced on previous day (MJ day^{-1})
ELY	Contribution of milk to the diet of the lamb (MJ day^{-1})
EM	ME requirements for maintenance (MJ day^{-1})
EW	NE requirements for walking and grazing (MJ day^{-1})
EX	ME intake available for body weight gain (MJ day^{-1})
F(L,J)	Weight of fodder in class L (kg DM ha^{-1})
IG(I)	Array for graphing output
IK(K)	Status of stock class K
KY(K)	Class to which the lambs from ewes in class K belong
QK	Mean digestibility of the diet, excluding milk
QM(M)	Digestibility of herbage diet eaten by mob M
RD(L,J)	Relative intake achievable from fodder class L in unit J
RD(6,J)	Relative intake achievable from unit J
RF,RL	Proportion of lamb's diet from fodder and milk respectively
SAREA	Relative surface area of animal
S(K)	Number of stock in class K
S1(K)	Proportion of stock with single offspring
S2(K)	Proportion of stock with twins
SM(M)	Number of stock in mob M
SR	Number of stock per ha in a paddock
SS	Proportion of stock dying from undernutrition
TKK	Function of lamb age (days)
TYR	Current time of year, counted from 1 January (days)
UG	Efficiency of use of ME for weight gain
UI	Age factor for calculating basal metabolism
UM	Efficiency of use of ME for maintenance
W(K)	Weight of clean wool on each sheep in class K (kg)

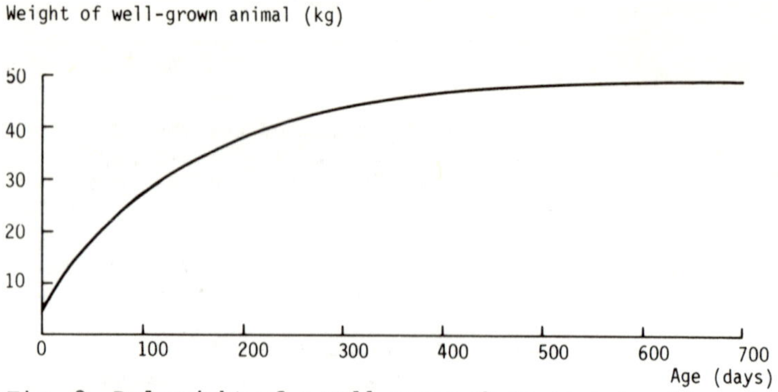

Fig. 9. Bodyweight of a well-grown sheep as a function of its age (days), using the equation, weight (kg)=50 - $45e^{-0.007(age)}$

than an exponential one:

$$y = ui\frac{n}{a}(2 - \frac{n}{a})$$

where

y = potential intake of dry matter (kg day^{-1})
u = upper limit of y for a non-lactating animal (kg day^{-1})
i = factor for lactation (calculated on the previous day at line 69)
n = normal body weight (kg)
a = mature weight (upper limit of normal body weight) (kg)

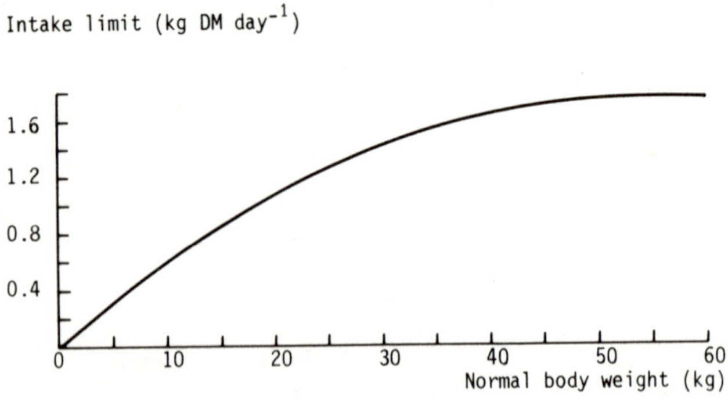

Fig. 10. Upper limit to the daily intake of food dry matter by a sheep, as a function of its normal bodyweight.

The calculations of actual intake and energy balance between lines 27 and 46 are affected by the contribution of milk or hay to the diet. The contribution of milk to the intake of ME is identified in line 27 as ELY and this has a positive value only for unweaned lambs, where EL(K) represents the intake of ME per lamb from milk. For their mothers, EL(K) has a negative value because the ME content of the milk produced is debited to the ewes' energy balance; for all other classes EL(K) is zero. At line 28, DK(K) is reduced by the herbage dry matter equivalent of the milk consumed. The remainder represents the upper limit to dry matter intake by a sheep eating a herbage diet of digestibility 0.8. These values are summed as DM(3,M) for all animals in each mob (line 29) for later adjustment of herbage weight.

The dry matter intake of herbage per animal, DF, is calculated at line 30 from DK(K) and the mean relative intake for all paddocks grazed by mob M, DM(2,M)/DM(1,M) (see FODDER). The total intake for all animals is accumulated as DFT (line 31). Dry matter intake of hay per animal in each mob, DHAY, is obtained (line 32) by multiplying the potential intake by DM(4,M), which is calculated in subroutine DRAFT. The total amount of hay eaten is accumulated as DHAYT (line 33). The intake of digestible dry matter, D(K), is the sum of herbage and hay intakes multiplied by the respective digestibilities QM(M) and C(19) (line 34). The mean digestibility of the diet, excluding milk, is QK (line 35). In the next line, D(K) is converted to metabolizable energy and the ME consumed as milk is added to give the total ME intake E(K) in MJ day^{-1}. RL and RF indicate the proportion of the ME intake derived from milk and herbage respectively; RF = 1.0 except for suckling lambs. Lines 39 and 40 calculate the efficiency of the use of ME for maintenance, UM, and for body weight gain, UG. The factors for the herbage part of the diet are taken from A.R.C. (1965) and those for the milk component from Walker and Norton (1971). The requirements for maintenance are calculated in lines 41-44. Basal metabolism, EB, is derived from the following equation of Graham, Searle and Griffith (1974), illustrated in Fig. 11:

$$h = 0.257\, w^{0.75} e^{-0.00022t} + 0.056\, m + 2.8\, g$$

where

h = basal metabolic rate (MJ day^{-1})
w = body weight (kg)
t = age (days)
m = intake of ME (MJ day^{-1})

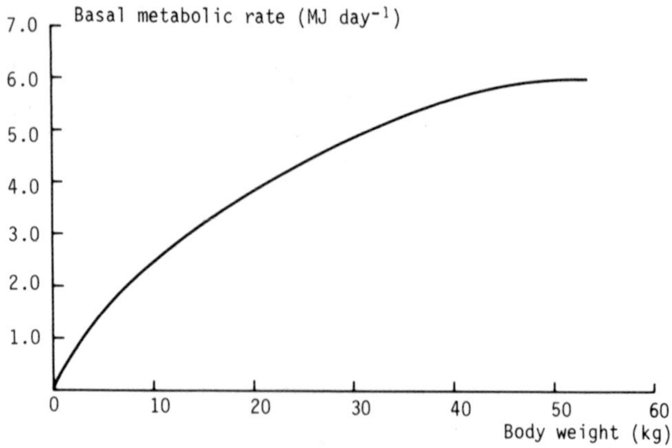

Fig. 11. Basal metabolic rate of a sheep in relation to its bodyweight, assuming that the sheep is well-grown for its age and that intake is at the upper limit.

g = body weight gain (kg day^{-1}).

To EB is added the energy required for walking and grazing, EW, calculated from Graham (1964b) as a function of body weight and predicted activity to give the total net energy required for maintenance and this is converted by UM to the ME requirement at line 44.

After subtracting this value from the ME intake, the remainder, EX, is corrected for the effect of feeding level on the ME content of the herbage component of the diet (line 46). There is room for doubt that this effect is significant (Graham 1961), but we have used the following function (Fig. 12) derived from the one recommended by A.R.C. (1965):

$$x_c = x(1 + 0.11f(\frac{1}{q} - 1))$$

where

x = ME intake surplus to maintenance requirements (MJ)
x_c = corrected value of x (MJ)
f = (intake of ME from herbage)/(maintenance requirements)
q = digestibility of herbage diet

If the ewes are pregnant (i.e. KK > 0 and AGE(KK) < 0) the routine calculates the weight of single and twin foetuses up to birth and the metabolizable energy required by the ewe for

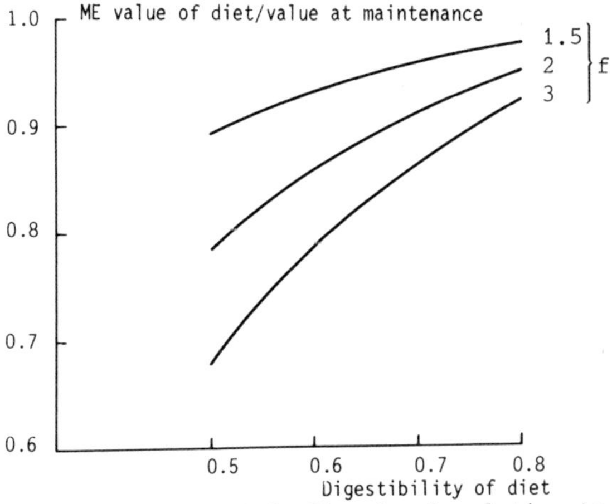

Fig. 12. Effect of feeding level relative to maintenance (f) and of digestibility on the metabolizable energy value of the fodder component of the diet

the development of the conceptus and the mammary gland. The daily weight gain of the foetus depends not only on the time from mating, TKK (line 51), but also on the condition of the ewe. This latter dependence increases as gestation progresses and is considerably greater for the ewe carrying twins. In line 52 the daily weight gain of a single foetus in a ewe of normal mature weight is calculated as BGY from the following function derived from Langlands and Sutherland (1968):

$$g = n10^{-6}(182.3 - 13.95t + 0.2057t^2)$$

where

g = daily gain in weight of a single foetus (kg)
n = normal body weight of ewe (kg)
t = time from mid-mating (days)

In the following line this value is accumulated daily as the normal weight of a single foetus, BN(KK). Normal gain is then adjusted for the condition of the ewe, to give the actual daily weight gain of a single lamb, BG1. If, for example, the ewe's weight is 20% below the normal mature weight, then the depression in the daily gain of a single foetus will increase from zero to 20% during gestation. The average reduction would be 10%, which is consistent with the results of Russel et al. (1967). Daily gain is accumulated to give the actual weight of the single foetus, BI(K), in line 55.

The daily weight gain of a twin foetus, BG2, is calculated from BG1 (line 56), using the assumptions that if the ewe is in good condition the gain by the foetus is 0.85 of that achieved by a single foetus, and that if the ewe's weight is less than the normal mature weight, the depression in the daily gain of a twin foetus is 2.5 times that of a single foetus (Russel et al. 1967). In line 57, BG2 is accumulated as the weight of a twin foetus, B2(K), and in the following line the mean weight of all foetuses from ewes in class K is calculated as B(KK). The values predicted from this routine for the weight of single and twin lambs at birth are consistent with the results of Curll et al. (1975).

The metabolizable energy required daily for the development of the whole conceptus in a ewe of normal mature weight with a single foetus, EY, is calculated at line 59 from the following function which is derived from Langlands and Sutherland (1968) and is illustrated in Fig. 13:

$$c = n10^{-6}(4178.0 - 295.2t + 4.788t^2)$$

where

c = ME required for development of conceptus (MJ day^{-1})
n = normal body weight of ewe (kg)
t = time from mid-mating (days)

During the last 42 days of gestation, the predicted requirement for mammary gland development is added to EY (line 60). This increment is calculated as 0.02 times the number of days after the fifteenth week of gestation, a function derived from the the results of Rattray et al. (1974), after scaling for the size of ewe. As the total requirement is for a ewe in normal condition with a single foetus, it is then adjusted (a) according to the proportions of single- and twin-bearing ewes and (b) to the weights of single and twin foetuses relative to the weight of a single foetus from a ewe in normal condition (line 60). This mean requirement for all ewes in class K is then subtracted from the value of EX before the weight change of the ewe is calculated. In all calculations concerning the pregnant ewe, her body weight does not include the weight of the conceptus or the increase in the weight of the udder.

After lambing occurs (AGE (KK) > 0) control, line 50, is directed to line 62 and the mean potential milk output per

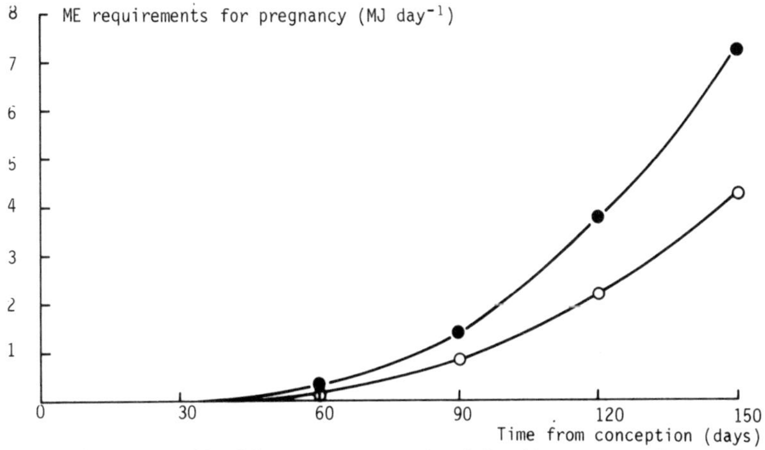

Fig. 13. Metabolizable energy required by the pregnant ewe for the development of the conceptus and the mammary gland.
○——○, ewe of normal mature weight with single foetus;
●——●, ewe of normal mature weight with twin foetuses.

ewe, ELPOT, is calculated (line 64) by the following function (Fig. 14). This is derived from the general equation for the lactation curve by Wood (1969) but modified for the effects of animal size, the condition of the ewe at lambing (calculated as BO(K) at line 106 of REPRO) and the proportion of twins (Davies 1963, Corbett 1968).

$$s = 0.05\ a^{0.75}\ t\ e^{-0.045t} (2\frac{w}{a} - 1)\ (1 + 0.5p)$$

where

s = potential output of milk ME per ewe (MJ day^{-1})
w = weight of ewe at lambing (kg)
t = time from lambing + 4 (days)
p = proportion of ewes with twins
a = mature weight.

If the ME intake by the ewe is less than that required to produce the potential output of milk the latter is reduced by a proportion, ELB, of the difference between the two, the proportion increasing with stage of lactation (line 65). In this calculation of the actual milk output (line 66) it is assumed that the ewe's intake of ME surplus to maintenance is converted to milk ME with an efficiency of 0.7 (A.R.C. 1965). Once the peak of the lactation curve has passed, the actual

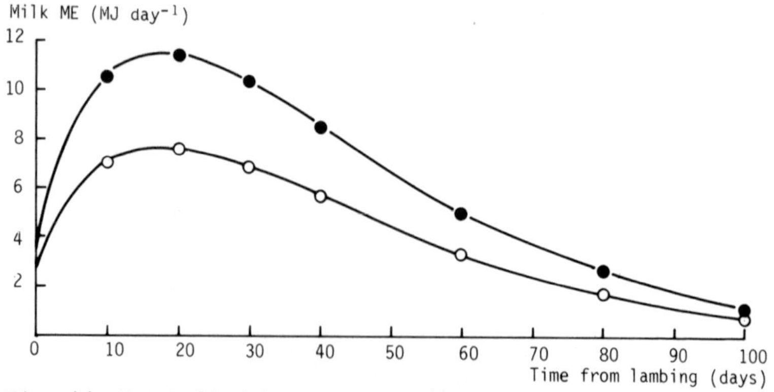

Fig. 14. Metabolizable energy content of the potential daily milk production of the ewe of normal mature bodyweight. O——O, ewe with single lamb; ●——●, ewe with twin lambs.

milk yield is prevented at line 67 from rising higher than the output on the previous day, ELMAX, even if there is an improvement in the ewe's intake of food. This possibility becomes less likely as lactation progresses, because potential intake follows a similar pattern to the lactation curve, but with a peak at 28 rather than 18 days after parturition. The daily changes in potential intake are effected by the intake factor DI(K) which is calculated in lines 70 and 71 from the following function (Fig. 15). This is based on the results of Hadjipieris and Holmes (1966) and Arnold and Dudzinski (1967).

$$i = 1 + 0.017\, t^{1.4}\, e^{-0.05t}\, (1 + 0.5p)$$

where

i = factor by which potential intake by ewe is multiplied during lactation
t = time from mid-lambing (days);
p: as above

Line 72 ensures that lambs less than 3 weeks old do not eat herbage, even if their milk consumption is low, since their intake factor, DI(KK), is almost zero during this time.

At line 68 the milk consumption per lamb, EL(KK), is calculated (as a positive quantity) from the milk output per ewe after adjusting for the proportion of twins. The ME required by the ewes to produce this milk is deducted from the previous value of EX (line 69) but the maintenance requirement is not adjusted

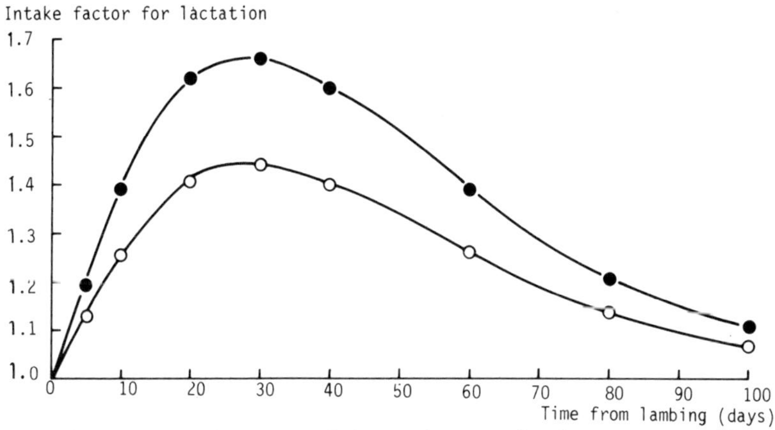

Fig. 15. Effect of stage of lactation on the factor controlling the potential intake of food by the ewe. Symbols as in Fig. 14.

for lactation as there appears to be no evidence of any increase other than that associated with increased intake (Graham 1964a).
The residual value of EX for all classes of stock is converted at lines 73 to 75 to body weight change, BG(K), by the following relationship (Fig. 16) derived from the results of Searle and Graham (1970).

$$g = s/(1.37 + 29.2w)$$

where

g = weight gain (kg day^{-1})
s = net energy intake surplus to maintenance (MJ day^{-1})
w = body weight/mature body weight

The weight change is added to body weight at line 76.

The weight of clean wool, W(K), on the sheep (which is not included in the computed body weight) is incremented daily at lines 77-79. During body development the relative density of wool follicles, AGEW, declines exponentially (line 77) and the relative surface area, SAREA, of the animal increases (line 79) (Schinckel and Short, 1961). These two factors and the effect of ME intake (other than that used for lactation)(Ferguson, 1962, 1970) are included in the function (Fig. 17) for daily wool growth at line 79:

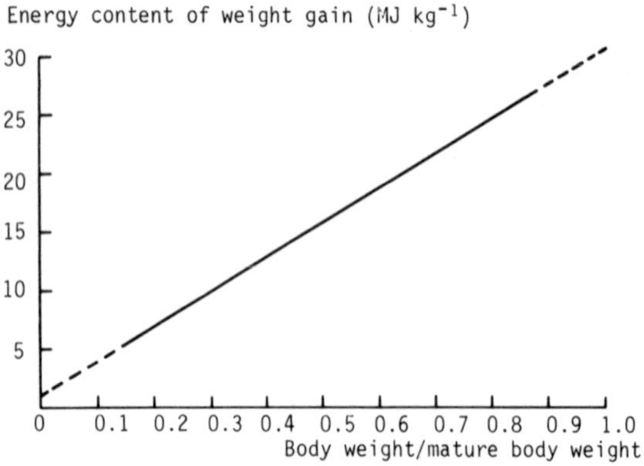

Fig. 16. Gross energy content of bodyweight gain in relation to the stage of development of the sheep.

$$v = 0.001(1 + e^{-0.011t})\left(\frac{w}{a}\right)^{0.67} m(1.5 - 0.037m)$$

where

v = daily growth of clean wool (kg)
t = age (days)
w = body weight (kg)
a = mature body weight (kg)
m = intake of metabolizable energy, less that used for lactation (MJ day^{-1}).

Lines 80-83 calculate the effect of undernutrition on the death rate of sheep, using unpublished results of Morley and Donnelly. A critical body weight, BC, is calculated at line 80 as a function of the stage of development of the stock in class K. The proportion of sheep dying increases by 0.003 for each 1% by which body weight falls below the critical body weight. The number and weight of surviving sheep are adjusted in lines 82 and 83, on the assumption that the sheep that died were 10% lighter than the mean of the surviving sheep. The number of sheep in each mob is calculated (line 84) and the age of all sheep is incremented by one day (line 87).

In lines 89 to 99 the herbage eaten by the animals and the dead herbage trampled is deducted from the appropriate herbage classes in each paddock. To distribute these amounts among the various paddocks grazed simultaneously by a mob it is necessary to calculate the effective number of stock per ha, SR, on each

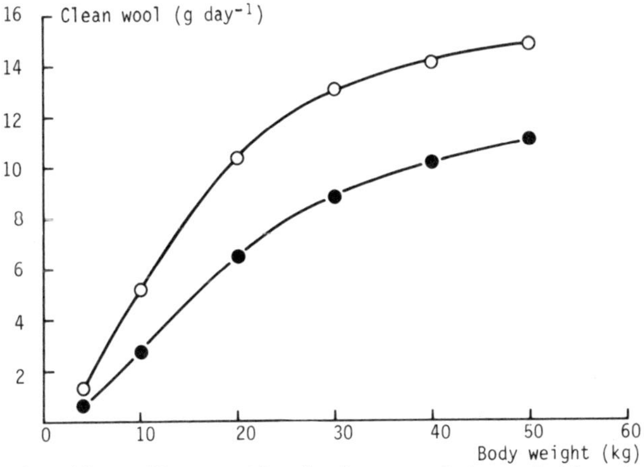

Fig. 17. Daily growth of clean wool in relation to the body-weight of the sheep at two levels of food intake, assuming that the sheep is well-grown for its age. ○——○, intake at the upper limit; ●——●, intake at 50% of the upper limit.

paddock. This is done at line 92 by assuming that the sheep are distributed between the paddocks in proportion to the product of the relative intake achievable from each paddock and the area of each paddock.

i.e. No. of sheep in paddock J = SM(M)* RD(6,J)* A(J)/DM(1,M)
∴ No. of sheep ha^{-1} in paddock J = SM(M)* RD(6,J)/DM(1,M)

The total potential intake of dry matter by the sheep in mob M, DM(3,M) is then distributed between the paddocks in proportion to SR to give the potential intake per ha from a paddock, DJJ (line 93). A proportion of the dead herbage (0.001 per sheep ha^{-1} day^{-1}) is deducted to account for trampling losses (line 94) but obviously these losses will be affected by the weather and the nature of the herbage. The grazed herbage is then subtracted from each herbage class (line 97) according to DJJ and the relative intake for each class, RD(L,J). Finally the herbage classes are summed to give a new total weight of dry matter per ha on the paddock (line 98).

The amount of hay in each digestibility class, F(L,1) is reduced by the amount eaten in lines 101-104. If stocks of hay fall to a predetermined level, C(20), subroutine MEMO is called to arrange purchase (line 105).

Lines 106 to the end enable the weights of herbage and stock to be plotted at chosen intervals during the year. These options are controlled by the values given to TLU(4) and TLU(5).

Program subroutine STOCK

```
          SUBROUTINE STOCK
          COMMON/A/RD(6,20),DM(4,10),FM(10),SM(10),QM(10),TT,TYR,DFT
          COMMON/B/T(20,6),F(6,20),FL(5,3),Q(5),CFG(27,3),FG(3),TLU(10)
          COMMON/C/C(100),A(20),S(20),B(20),BN(20),AGE(20),W(20),
     *    DI(20),EL(20),PJ(20),PK(20),VJ(20),VK(20),FI(20),TI(20),TK(20)
          COMMON/D/D(20),DK(20),E(20),BG(20),MJ(20),MK(20),NT(20,6)
          COMMON/Y/S0(2),S1(2),S2(2),B1(2),B2(2),B0(2),TMID(2)
          COMMON/O/OJ(20),OK(20),OW(20),OV(20),XOP(40),COP(20),DOP(20),
     *    IOP(20),NOP,NRUN,NDO,IYR,IO,IRUN,IDO,IREV,IMP,OO,OOO,OEI,OEIT
          COMMON/M/NFG(20),IJ(20),IK(20),KRE(20),KY(20)
          COMMON/N/NJ,NK,NL,NM,NP,NDP,NYR,LU,LFG,L1,L2,L3,L4,L5,L6,L7,L8
          COMMON/G/IIG,IG(125),LG(8),LO,LDOT
 101      FORMAT(F5.0,11F7.2,8X,8F5.1)
 111      FORMAT(F5.0,125A1)
 112      FORMAT(F5.0,125A1)
          DFT=0.0
          DHAYT=0.0
          DO  60 K=1,NK
          IF(S(K).LT.1.0) GO TO 60
          IF(AGE(K).LT.1.0) GO TO 59
          I=IK(K)
          KK=KY(K)
          M=MK(K)
          BNMAX=BN(20)-(BN(20)-C(22))*EXP(-.01*C(23)*AGE(K))
          BN(K)=AMIN1(AMAX1(B(K),BN(K)),BNMAX)
          DK(K)=C(24)*DI(K)*(BN(K)/BN(20))*(2.0-BN(K)/BN(20))
          ELY=AMAX1(EL(K),0.001)
          DK(K)=DIM(DK(K),ELY/(C(25)*Q(5)))+0.000001
          DM(3,M)=DM(3,M)+S(K)*DK(K)
          DF=DK(K)*DM(2,M)/DM(1,M)
          DFT=DFT+DF*S(K)
          DHAY=DM(4,M)*DK(K)
          DHAYT=DHAYT+S(K)*DHAY
          D(K)=QM(M)*DF+C(19)*DHAY
          QK=D(K)/(DF+DHAY)
          E(K)=C(25)*D(K)+ELY
          RL=ELY/E(K)
          RF=1.0-RL
          UM=(0.546+0.246*QK)*RF+0.84*RL
          UG=(0.03+0.664*QK)*RF+0.69*RL
          UI=EXP(-.00022*AGE(K))
          EB=(C(26)*UI*B(K)**0.75+C(27)*E(K)+C(28)*BG(K))*(1.0+.25*RL)
          EW=B(K)*(C(29)*RF+C(30))
          EM=(EB+EW)/UM
          EX=E(K)-EM
          IF(EX.GT.ELY)EX=EX*(1.0-((E(K)-ELY)/EM)*(C(32)*(1.0/QK-1.0)))
          DI(K)=1.0
          IF(KK.LE.0) GO TO 40
          S(KK)=S(K)*(S1(K)+2.0*S2(K))
          IF(AGE(KK).GT.0.0) GO TO 31
          TKK=AGE(KK)+C(97)
          BGY=BN(K)*1.0E-6*(182.3+TKK*(-13.95+TKK*0.2057))
          BN(KK)=BN(KK)+BGY
          BG1=BGY*(B1(K)*B(K)/BN(20)+C(22)-B1(K))/C(22)
          B1(K)=B1(K)+BG1
          BG2=AMIN1(0.85*BG1,0.85*BGY-2.5*(BGY-BG1))
```

```
      B2(K)=B2(K)+BG2
      B(KK)=(S1(K)*B1(K)+2.0*S2(K)*B2(K))/(S1(K)+2.0*S2(K))
      EY=BN(K)*1.0E-6*(4178.0+TKK*(-295.2+TKK*4.788))
      EX=EX-(EY+0.02*DIM(TI(5),-AGE(KK)))*(S(KK)*B(KK))/(S(K)*BN(KK))
      GO TO 40
   31 ELMAX=EL(K)
      TKK=AGE(KK)+4.0
      ELPOT=BO(K)*TKK*EXP(-C(35)*TKK)*(S1(K)+1.5*S2(K))
      ELB=0.5/(1.0+0.9E5*EXP(-0.38*AGE(KK)))
      EL(K)=-(ELPOT-ELB*DIM(ELPOT,C(36)*EX))
      IF(AGE(KK).GT.1.0/C(35)) EL(K)=AMAX1(ELMAX,EL(K))
      EL(KK)=-EL(K)*S(K)/S(KK)
      EX=EX+EL(K)/C(36)
      DI(K)=C(37)*AGE(KK)**C(38)*EXP(-C(39)*AGE(KK))
      DI(K)=1.0+DI(K)*(S1(K)+1.5*S2(K))*(2.0*B(K)/BN(20)-1.0)
      DI(KK)=1.0/(1.0+5020.0*EXP(-.31*AGE(KK)))
   40 IF(EX.GT.0.0) EX=UG*EX
      IF(EX.LT.0.0) EX=C(41)*EX
      BG(K)=EX/(C(42)+C(43)*B(K)/BN(20))
      B(K)=B(K)+BG(K)
      AGEW=1.0+EXP(-C(44)*AGE(K))
      SAREA=0.001*AMIN1(1.0,(B(K)/BN(20))**0.67)
      W(K)=W(K)+SAREA*AGEW*(EX+EM)*(C(45)-C(46)*(EX+EM))
      BC=BN(K)*(1.0-C(49)*(1.0+BN(K)/BN(20)))
      SS=S(K)*(0.001*C(47)+C(48)*DIM(1.0,B(K)/BC))
      B(K)=B(K)*S(K)/(S(K)-SS*(1.0-C(50)))
      S(K)=S(K)-SS
      SM(M)=SM(M)+S(K)
      IF(K.EQ.1.AND.AMOD(TT,TLU(3)).LT.0.9) WRITE(L3,101) TT,T(K,5),
     * DK(K),DF,DHAY,D(K),QK,E(K),EL(K),EX,BG(K),B(K),(D(L),L=1,8)
   59 AGE(K)=AGE(K)+1.0
   60 CONTINUE
      DO 70 J=1,NJ
      M=MJ(J)
      IF(M.EQ.0) GO TO 70
      SR=SM(M)*RD(6,J)/DM(1,M)
      DJJ=RD(6,J)*DM(3,M)/DM(1,M)
      F(1,J)=F(1,J)*(1.0-0.001*C(10)*SR)
      F(6,J)=0.0
      DO 65 L=1,NL
      F(L,J)=F(L,J)-DJJ*RD(L,J)
   65 F(6,J)=F(6,J)+F(L,J)
   70 CONTINUE
      CALL MEMO (1,1,TT+1.0)
      IF(DHAYT.LE.0.0) GO TO 81
      HX=1.0-DHAYT/(F(6,1)*A(1))
      DO 80 L=1,6
   80 F(L,1)=F(L,1)*HX
      IF(F(6,1).LT.C(20)) CALL MEMO(1,2,TT)
   81 IF (AMOD(TT,TLU(4)).GT.0.9) GO TO 91
      DO 82 I=1,125
   82 IG(I)=LO
      DO 84 I=20,100,20
   84 IG(I)=LDOT
      DO 85 L=1,5
      F(L+1,20)=F(L+1,20)+F(L,20)
      I=MINO(MAX1(0.02*F(L+1,20),0.0),125)
   85 IG(I)=LG(L)
      WRITE(L4,111) TT,(IG(L),L=1,I)
   91 IF(AMOD(TT,TLU(5)).GT.0.9) GO TO 100
      DO 92 I=1,125
   92 IG(I)=LO
      DO 94 I=20,100,20
   94 IG(I)=LDOT
      DO 95 K=1,8
      I=MINO(MAX1(2.0*B(K),0.0),125)
   95 IG(I)=LG(K)
      WRITE(L5,112) TT,IG
  100 RETURN
      END
```

8 Subroutine RATION

This subroutine deals with changes in the status of fodder units, particularly those changes relating to haymaking. The various changes possible are shown diagrammatically in Fig. 18. Only four types of unit are recognized in the model at present (Table 10), although paddocks of different fertility or plant species, crops or supplementary feeds of various kinds could be added to the list where applicable. The variables used in the subroutines are defined in Table 11.

At the start of each run, those paddocks which may be cut for hay are designated by setting the times for the first event T(J,2) at the earliest date, C(13), at which closing up paddocks for hay is first considered.
A paddock is closed for deferred grazing or for haymaking merely by changing the status from 1 to 2 or to 3 respectively, and opened again by the reverse change. Since the priority values for units with status 2 and 3 are normally high, they automatically remain unavailable except during feed shortage (this procedure is explained more fully in Section 10 on DRAFT). Although haymaking is confined to a definite period during the season, grazing may be deferred at any time of the year, depending on the feed situation. Accordingly, changes dealing with deferred grazing are more conveniently carried out in subroutine DRAFT.

The first step in subroutine RATION is to find out from the second row of the NT array which fodder unit, J, has the next scheduled event (line 13), so that control can be directed to

Table 10 Status levels for fodder units

Status	Description
1	Paddock available for grazing
2	Paddock on which grazing is deferred
3	Paddock shut up for hay
4	Hay store

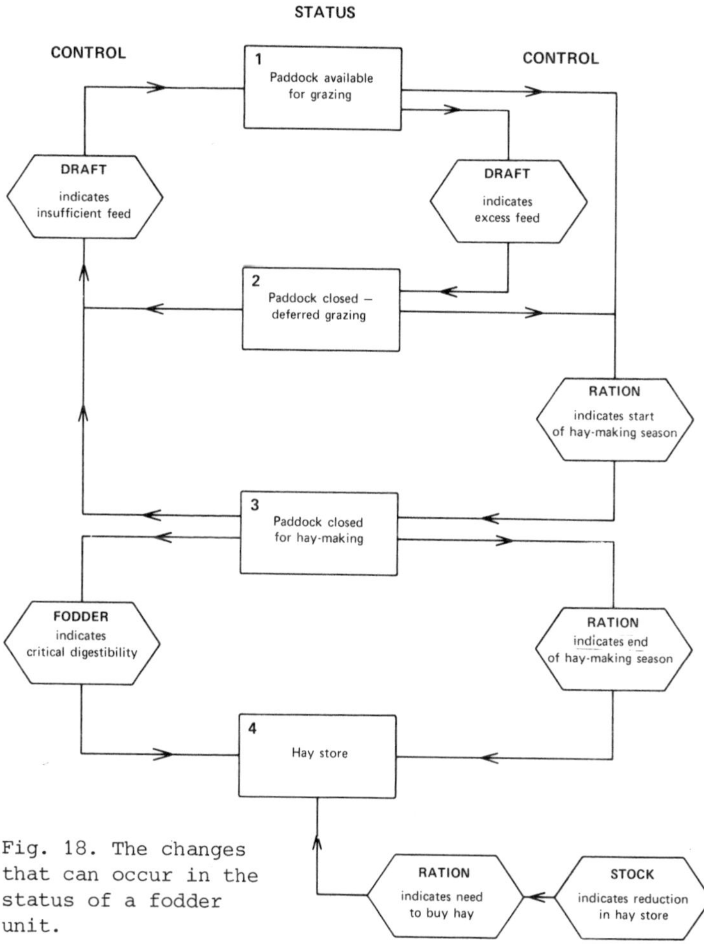

Fig. 18. The changes that can occur in the status of a fodder unit.

the segment of the routine which deals with the event appropriate to the status of that unit (line 16).
If the unit is a paddock which is either available for grazing or being saved for later grazing (status 1 or 2), the scheduled event will be the closing of the paddock for hay. However, if the current date, TT, is later than 50 days before the end of the hay-making season, C(14), the paddock is, instead, rescheduled to be closed for hay in the following year.

Table 11 Definition of variables in subroutine RATION

Variable name	Definition
F(L,J)	Weight of fodder in class L of unit J (kg DM ha^{-1})
F(6,J)	Total weight of fodder in unit J (kg DM ha^{-1})
F(6,1)	Weight of hay in store (tonnes DM)
FH	Hay conserved from paddock (tonnes DM)
FQ	Digestible DM in hay (tonnes)
FX	Herbage DM removed as hay (kg ha^{-1})
FXX	Hay conserved from each class of herbage in paddock (kg ha^{-1})
OJ(1)	Value of hay in store ($)
Q(L)	Digestibility of fodder in class L
TT	Current date
VJ(4)	Purchase price of hay ($$kg^{-1}$)

(line 17).

At harvest, the status of the paddock is restored to 1 (line 21). The weight of material in classes with digestibility 0.6 - 0.8 is set to zero, but it is assumed that for the fodder of digestibility 0.4 and 0.5 only amounts in excess of 500 kg/ha can be removed, the rest remaining as stubble (lines 25-27). It is further assumed that a constant fraction C(15) of each class is conserved in the haymaking process, and this amount is transferred to fodder unit 1 (lines 29-30), which for convenience is designated as the hay store, with status 4. In this listing MEMO then schedules the same paddock to be shut up for hay again at the same time next year (line 37) but a more flexible policy could be adopted.
The cost of harvesting is debited according to the amount of hay dry matter (line 36), and the mean digestibility of the hay in unit 1, C(19), is calculated at line 35.
If the fodder unit is the hay store (I = 4), the call indicates that the amount present, as calculated in subroutine STOCK, has fallen below a stipulated amount. Accordingly, the same amount, C(20), is purchased at VJ(4) dollars per kg, and an assumed digestibility of 60% (digestibility class 3, line 39). The mean digestibility of the hay in the unit is then recalculated (line 46).

```
      SUBROUTINE RATION
      COMMON/A/RD(6,20),DM(4,10),FM(10),SM(10),QM(10),TT,TYR,DFT
      COMMON/B/T(20,6),F(6,20),FL(5,3),Q(5),CFG(27,3),FG(3),TLU(10)
      COMMON/C/C(100),A(20),S(20),B(20),BN(20),AGE(20),W(20),
     * DI(20),EL(20),PJ(20),PK(20),VJ(20),VK(20),FI(20),TI(20),TK(20)
      COMMON/D/D(20),DK(20),E(20),BG(20),MJ(20),MK(20),NT(20,6)
      COMMON/Y/S0(2),S1(2),S2(2),B1(2),B2(2),B0(2),TMID(2)
      COMMON/O/OJ(20),OK(20),OW(20),OV(20),XOP(40),COP(20),DOP(20),
     * IOP(20),NOP,NRUN,NDO,IYR,IO,IRUN,IDO,IREV,IMP,OO,OOO,OEI,OEIT
      COMMON/M/NFG(20),IJ(20),IK(20),KRE(20),KY(20)
      COMMON/N/NJ,NK,NL,NM,NP,NDP,NYR,LU,LFG,L1,L2,L3,L4,L5,L6,L7,L8
  101 FORMAT(F5.0,2I5,9X,6F7.0)
      J=NT(20,2)
      I=IJ(J)
      IF(TLU(6).GT.0.0) WRITE(L6,101) TT,J,I,(F(L,J),L=1,6)
      GOTO(11,11,31,41,99,99,99,99,99,99,99,99,99,99,99,99,99,99,99)I
   11 IF(TT.GT.C(14)-50.0) GO TO 36
      IJ(J)=3
      CALL MEMO (J,2,C(14))
      GO TO 99
   31 IJ(J)=1
      FQ=0.0
      FH=0.0
      DO 35 L=1,NL
      FX=F(L,J)
      IF (L.LT.3) FX=DIM(FX,500.0)
      F(L,J)=F(L,J)-FX
      F(6,J)=F(6,J)-FX
      FXX=C(15)*FX*A(J)/A(1)
      F(L,1)=F(L,1)+FXX
      FQ=FQ+Q(L)*F(L,1)
      FH=FH+FXX
   35 CONTINUE
      F(6,1)=F(6,1)+FH
      C(19)=FQ/F(6,1)
      OJ(J)=OJ(J)-C(79)*FH*A(1)
   36 CALL MEMO (J,2,C(13)+365.0)
      GO TO 99
   41 F(3,1)=F(3,1)+C(20)
      OJ(1)=OJ(1)-VJ(4)*A(1)*C(20)
      FQ=0.0
      F(6,1)=0.0
      DO 45 L=1,NL
      FQ=FQ+Q(L)*F(L,1)
   45 F(6,1)=F(6,1)+F(L,1)
      C(19)=FQ/F(6,1)
      CALL MEMO (J,2,1000.0)
      GO TO 99
   99 CALL MEMO (4,1,T(1,1)-0.1)
      WRITE(L6,101) TT,J,IJ(J),(F(L,J),L=1,6)
      RETURN
      END
```

9 Subroutine REPRO

This routine calculates the effects on stock class attributes of various management parameters and physiological changes, in particular those relating to the reproductive cycle. The variables used in the subroutine are defined in Table 12. As shown in Fig. 19, the program is broken up into a number of segments, each dealing with a particular stock event. For example stock events such as the start and end of mating involve changes in the status of a stock class because priorities change and new mobs must be drafted, but during mating other stock events will occur which merely require calculations associated with the individual oestrus cycles and do not involve a change of status. It is important to note that only one segment of the program is executed during each call to the subroutine.

The flag IDRAFT is initially set to zero (line 14) to indicate that a need has not yet been established for redrafting the sheep into new mobs. The stock class K to be dealt with is that with the lowest T(K,3) value (i.e. current time), as indicated by its position in the NT array (line 15). Selection of the segment appropriate to a particular call depends not only on the current status of the stock class but also on whether the stock class requires a change of status. If it does not, i.e. if the time interval TI(IK(K)) between status changes has not expired (line 20), then the statement number to which control is transferred in line 27 is determined by the unchanged status (line 26). However, if a change in status is due, this is incremented by one, the interval counter TK(K) is restored to zero, the flag is set for re-drafting and the control index, IX, is calculated from the new status (lines 21-24).
Since both mature and maiden ewes follow the same sequence of status changes in fixed intervals from flushing to weaning, it is convenient for all sheep in the same physiological condition to have status values which always differ by a constant. For example (Fig. 3) mature and maiden ewes at flushing will have status 2 and 12 respectively, while at lambing these are 6 and 16 respectively. This results in the same value of IX applying to both classes of sheep (lines 24

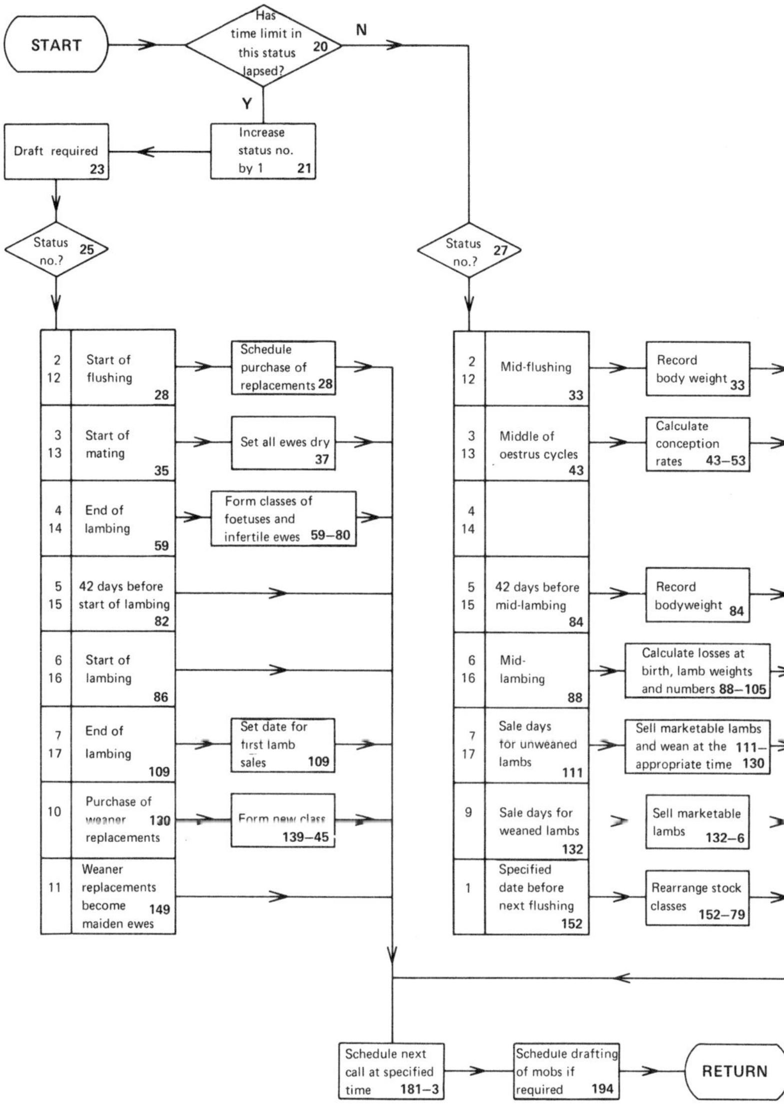

Fig. 19. Flow chart for subroutine REPRO.

Table 12 Definition of variables in subroutine REPRO

Variable name	Definition
A0	Probability of no ovulations
A1	Probability of 1 ovulation
A2	Probability of 2 ovulations
AGE(K)	Age of animals in class K (days)
B(K)	Body weight of animals in class K (kg)
B1(K)	Weight of a single foetus in ewe in class K (kg)
B2(K)	Weight of a twin foetus in ewe in class K (kg)
BGF	Function of weight gain before mating
BN(K)	Normal body weight of animals in class K (kg)
BO(K)	Function of body weight at a specified time
CC	Seasonal factor for fertility
DK(K)	See STOCK
EL(K)	See STOCK
K1	Class being transferred during reorganization at end of year
K2	Class to which K1 is transferred during stock reorganization
KRE(K)	List of classes to which transfer is made
KY(K)	Class of lambs belonging to ewe class K
IDRAFT	Flag indicating need for redrafting of mobs
IK(K)	Status of stock class K
IX	Index for directing program control
OK(K)	Objective function for stock class K
OSK	Factor used for calculating value of animals sold
OV(K)	Variable costs for stock class K ($)
PF	Probability of fertilization
PO	Probability of each ovulation
S(K)	No. of sheep in class K
S0(K)	Proportion of non-fertile ewes
S1(K)	Proportion of ewes in class K with one foetus or lamb
S2(K)	Proportion of ewes in class K with twin foetuses or lambs
SF	Proportion of fertile ewes
SS	Proportion of stock class sold, dying or transferred
TI(IK(K))	Time interval during which status remains constant (days)
TK(K)	Counter during each time interval TI (days)
TMID(K)	Interval between start of mating and mean date of conception (days)
TT	Current time (days)

Table 12 (continued)

Variable name	Definition
TX	Calculated time interval (days)
TYR	Time of year, from 1 January (days)
W(K)	Weight of clean wool in fleeces of sheep in class K (kg)
ZDYS	Proportion of single lambs dying (dystocia)
ZPTOX	Proportion of twin lambs dying (pregnancy toxaemia)
ZX	Factor for calculating losses from mismothering and exposure
ZXPOS1	Proportion of single lambs dying (mismothering and exposure)
ZXPOS2	Proportion of twin lambs dying (mismothering and exposure)

and 26).
Starting with dry mature ewes, where IK(K) = 1 and hence IX = 1, the first event is the start of flushing at the time T(1,3) read in at the start of the run; this also applies to maiden ewes with IK(K) = 11.
If the values of TK(1) and TI(1) that were read in at the start of the run are equal, then the status of the class is increased to 2 (line 21) and control is directed at line 25 to statement 10 where MEMO is called to schedule the purchase of replacement weaners C(87) days after the start of the year and the status of these animals (class 4) is set at 10.
The value of TX is set at 12 (line 31) and this becomes the new value of TK(1) (line 182), with the effect that the next call is scheduled 12 days later. As a precaution, after each of these procedures, lines 185 to 191 detect stock classes that belong to no mob and mobs that contain no stock classes.
At the end of each call, a call to DRAFT is scheduled if IDRAFT = 1 (line 194) and the condition of the particular class may be printed (line 196) for comparison with their condition at the start (line 17).

At the next call for class 1 (mid-flushing), the low value of TK(K) relative to TI(IK(K)) prevents a change in status and the only action taken is the retention of the weight of the ewes as BO(K) (line 33). TX is recalculated as the time remaining in the current status, 9 days, and the next call is scheduled 9 days later (line 177). When this occurs, the status is increased to 3 to indicate the start of mating. Control is directed to statement 15 where the time interval, TX, before the next stock event, i.e. the middle of the first oestrus cycle, is calculated as 8.5 days. This interval is retained as a negative value, TMID(K), for subsequent estimation of the mean mating date.

The next two calls are at the mid-points of the first two oestrus cycles. The status of the class is unchanged at 3 and control is directed to statement 16 where the probabilities of ovulation, PO, and fertilization, PF, are predicted. These functions were developed by Morley (pers. comm.) from the experimental data of Radford (1959), Watson and Radford (1966) and Killeen (1967) and were described in detail by White (1975). The predictors of PO and PF are body weight relative to normal mature weight, body weight gain over the previous 17 days and a function of the time of year, CC. The proportions of ewes with 0, 1 and 2 ovulations are estimated from simple binomial probabilities (lines 47-49) and the proportions of ewes with twins, S2(K), and non-pregnant ewes, S0(K) are calculated (lines 50 and 51). In the second oestrus cycle, the probabilities are calculated in the same way and apply to the ewes remaining infertile after the first cycle. A weighted mean mating date, TMID(K), is then calculated in days from the start of mating (line 54) and the end of mating is scheduled in 8.5 days time.
It is assumed that the mean body weight of the dry ewes is C(51)kg less than that of the remainder of the flock (line 59). Whether maiden or mature ewes, they are put into a separate class (class 3) when mating ends. The parameters of this class are then recalculated (lines 64-70).
Also at the end of mating, a new class is opened for the foetuses, KY(K) = KK, with age determined by TMID(K) (line 80). The initial level of milk production after lambing depends on the weight of the ewe six weeks before lambing (see STOCK). Therefore, six weeks before the mean date of lambing, when IK = 5, the body weight value is stored as BO(K) (line 84).

The mean lambing date is TMID(K) days after IX becomes 6, and when AGE(KK) = 0. Using the simplification that all lambs from the ewes in class K are born on this day, estimates are made of neonatal losses. The proportion of single lambs that die from dystocia, ZDYS, is assumed to depend on the extent to which the weights of both lambs and ewes exceed the respective normal body weights (line 88). The predicted losses match the results of Alexander et al. (1955) and Curll et al. (1975). The proportion of twin lambs lost through pregnancy toxaemia, ZPTOX, is predicted from a function of body weight loss over

the last six weeks of gestation (line 89) which is based on the results of Reid and Hinks (1962) and Morley and Donnelly (pers. comm.). Although pregnancy toxaemia may, under experimental conditions, affect a ewe with a single foetus, this does not, in our experience, occur in the field.
Ewe numbers and the proportion of ewes with singles and twins are recalculated accordingly (lines 90-92) before deducting losses from mismothering and exposure. These losses are predicted (lines 94 and 95) from the weight deficit of the single and twin lambs, relative to the normal weight of a single lamb, and from the factor ZX (line 93). This factor has a constant component, C(58), and a seasonal component which mirrors the probability of inclement weather (Obst and Day 1968). In our environment this function follows the pattern of daily minimum temperature. The routine then calculates the weights of the surviving single and twin lambs and the mean weight of all lambs in the class (line 100). The proportions of single and twin-bearing ewes are adjusted on the assumption that deaths of twin lambs from exposure remove only one of a pair, leaving a single lamb. The total number of surviving lambs is calculated as S(KK) (line 103) and their normal body weight set at the current mean weight. Finally, in line 105, a function of the ewes' weight is calculated for the prediction of potential milk yield (see subroutine STOCK).

The feed intake of lambs from this time on is calculated independently in subroutine STOCK, since B(KK) > 1, but as they are supplied with milk by the ewes they are always drafted into the same mob.
The sale of prime lambs is scheduled to start after C(92) days from the end of lambing (line 109) and at intervals of C(93) days thereafter. The proportion of the remaining animals sold at any one time, SS, is given as a function of body weight B(K) as follows (line 111):

if

B(K) < C(67) : SS = 0;

if

C(67) < B(K) < (C(67) + C(68)) : SS = (B(K) - C(67))/C(68);

if

(C(67) + C(68)) < B(K) : SS = 1.0.

The mean body weight of the lambs sold is assumed to be C(69) kg higher than of those retained, and the sale value, based on this weight, is added to the objective function (line 114). The status of the remaining animals does not change. Weaning takes place either when the calculated milk production per ewe falls to C(40) $MJday^{-1}$ or at 85 days after lambing, whichever is the earlier (line 116). The lamb class KK is then made completely independent by putting IK(KK) = 9 (line 119) and sales continue (line 132-137). The calculations for the ewe class dealing with lactation are terminated (line 118), their status restored to 1 and the flag set (line 125) for the redrafting of the mobs.

The next event for the mature ewe class is the start of flushing in the following year; the time for this is calculated by adding TI(1) or 160 days to the mean lambing date (line 130). The procedure then diverges for young and mature ewes. The young ewes require no further attention at this stage (line 128) but the start of flushing for their replacements, the ewe weaner class, must be scheduled at the same calendar date as for the maiden ewes of the previous year (line 127). This class of weaners (class 4) was purchased earlier in the year (line 139-145) and the status was updated via lines 149-150. For the older ewes, a return to the subroutine 60 days before the start of flushing is scheduled to allow reorganization after disposal of the main crop of prime lambs and before the start of the new season (line 152). Because of the creation of new classes during the year, regrouping is needed to restore the situation to one comparable to that obtaining at the start of the previous year. The KRE array, as read in, determines where each class is to be relocated (line 153).

Dry ewes and ewes which have weaned their lambs (classes 1-3) are combined into class 1, while younger breeding stock move down one step. Unsold lambs are transferred into class 6, from which they will be sold during the early part of the following year (see lines 132-137). These transfers involve recalculating parameter values (lines 162-166), and setting previous values to zero (lines 167-170). A rescheduling of events is also required (line 173).
The opportunity is also taken at this time to cull ewes from the main flock. To maintain a reasonably stable flock size, the number culled is made equal to the excess of ewes at this time, if any, over the number of mature ewes at the start of the run, C(81). The sale value of these animals according to their body weight is added to the objective function (line 178).

Program subroutine REPRO

```
      SUBROUTINE REPRO
      COMMON/A/RD(6,20),DM(4,10),FM(10),SM(10),QM(10),TT,TYR,DFT
      COMMON/B/T(20,6),F(6,20),FL(5,3),Q(5),CFG(27,3),FG(3),TLU(10)
      COMMON/C/C(100),A(20),S(20),B(20),BN(20),AGE(20),W(20),
     * DI(20),EL(20),PJ(20),PK(20),VJ(20),VK(20),FI(20),TI(20),TK(20)
      COMMON/D/D(20),DK(20),E(20),BG(20),MJ(20),MK(20),NT(20,6)
      COMMON/Y/S0(2),S1(2),S2(2),B1(2),B2(2),BO(2),TMID(2)
      COMMON/O/OJ(20),OK(20),OW(20),OV(20),XOP(40),COP(20),DOP(20),
     * IOP(20),NOP,NRUN,NDO,IYR,IO,IRUN,IDO,IREV,IMP,OO,OOO,OEI,OEIT
      COMMON/M/NFG(20),IJ(20),IK(20),KRE(20),KY(20)
      COMMON/N/NJ,NK,NL,NM,NP,NDP,NYR,LU,LFG,L1,L2,L3,L4,L5,L6,L7,L8
  101 FORMAT(F5.0,2I4,F6.0,F6.0,2X,8F6.0,6X,8F6.1)
  102 FORMAT(I5,12F8.2)
      IDRAFT=0
      K=NT(20,3)
      IF(TT.LT.TLU(7)) GO TO 2
      WRITE(L7,101)TT,K,IK(K),TK(K),AGE(K),(S(L),L=1,8),(B(L),L=1,8)
    2 KK=KY(K)
      TX=0.0
      IF(TK(K)+0.001.LT.TI(IK(K))) GO TO 3
      IK(K)=IK(K)+1
      TK(K)=0.0
      IDRAFT=1
      IX=MOD(IK(K)-1,10)+1
      GO TO(61,10,15,20,25,30,35,91,45,50) IX
    3 IX=MOD(IK(K)-1,10)+1
      GO TO(61,11,16,91,26,31,36,91,45,50) IX
   10 CALL MEMO (4,3,C(87))
      IK(4)=10
      TK(4)=0.0
      TX=TI(2)-0.5*C(91)
      GO TO 81
   11 BO(K)=B(K)
      GO TO 81
   15 B1(K)=0.0
      B2(K)=0.0
      S0(K)=1.0
      S1(K)=0.0
      S2(K)=0.0
      TMID(K)=-0.5*C(91)
      TX=-TMID(K)
      GO TO 81
   16 BGF=0.04*(B(K)-BO(K))+0.8
      CC=SIN((TYR+10.0-0.5*C(91))*3.1416/182.5)-1.0
      PO=(1.0-C(61)*(C(62)-B(K)/BN(20))**2)*BGF*(1.0+C(63)*CC)
      PF=(1.0-EXP(-C(64)*B(K)/BN(20)))*BGF*(1.0+C(65)*CC)
      A0=(1.0-PO)*(1.0-PO)
      A2=PO*PO*C(66)
      A1=1.0-A0-A2
      SF=S1(K)+S2(K)
      S2(K)=S2(K)+S0(K)*A2*PF/(A1+A2*(2.0-PF))
      S0(K)=S0(K)*(A0+(1.0-PF)*(A1+A2*(1.0-PF)))
      S1(K)=1.0-S0(K)-S2(K)
      TMID(K)=TMID(K)+C(91)*(S1(K)+S2(K)-SF)/(S1(K)+S2(K))
      WRITE(L7,102)K,BGF,CC,PO,PF,A0,A1,A2,S0(K),S1(K),S2(K),TMID(K)
      BO(K)=B(K)
      TX=AMIN1(C(91),TI(3)-TK(K))
      GO TO 81
   20 B(K)=B(K)+C(51)*S0(K)
      N=3
      IK(N)=1
      SS=S0(K)*S(K)
      S(K)=S(K)-SS
      S(N)=S(N)+SS
      SN=SS/S(N)
```

```
      B(N)=B(N)+SN*(B(K)-C(51)-B(N))
      BN(N)=BN(N)+SN*(BN(K)-BN(N))
      AGE(N)=AGE(N)+SN*(AGE(K)-AGE(N))
      W(N)=W(N)+SN*(W(K)-W(N))
      DK(N)=DK(N)+SN*(DK(K)-DK(N))
      S0(K)=0.0
      SF=S1(K)+S2(K)
      S1(K)=S1(K)/SF
      S2(K)=S2(K)/SF
      KY(K)=-KY(K)
      KK=KY(K)
      IK(KK)=8
      S(KK)=S(K)*(S1(K)+2.0*S2(K))
      TMID(K)=AINT(TMID(K))
      AGE(KK)=-TI(4)-TI(5)-TMID(K)
      GO TO 81
   25 TX=TMID(K)
      GO TO 81
   26 BO(K)=B(K)
      GO TO 81
   30 TX=TMID(K)
      GO TO 81
   31 ZDYS=C(52)*(DIM(B(K),BN(K))+C(53)*DIM(B1(K),BN(KK)))
      ZPTOX=0.01*C(54)*DIM(BO(K),B(K))**2.0
      S(K)=S(K)*(1.0-ZPTOX)
      S1(K)=S1(K)*(1.0-ZDYS)
      S2(K)=S2(K)*(1.0-ZPTOX)
      ZX=C(58)+C(59)*0.5*(COS(3.1416*(TYR/182.5-1.0))+1.0)
      ZXPOS1=ZX*(1.0+C(57)*(DIM(C(22),B1(K)))**2.0)
      ZXPOS2=ZX*(1.0+C(57)*(DIM(C(22),B2(K)))**2.0)
      B1(K)=B1(K)+ZXPOS1*C(60)
      B2(K)=B2(K)+ZXPOS2*C(60)
      S1(K)=S1(K)*(1.0-ZXPOS1)
      S2(K)=S2(K)*(1.0-ZXPOS2)
      B(KK)=(S1(K)*B1(K)+2.0*S2(K)*B2(K))/(S1(K)+2.0*S2(K))
      S1(K)=S1(K)+S2(K)*ZXPOS2
      S0(K)=1.0-S1(K)-S2(K)
      S(KK)=S(K)*(S1(K)+2.0*S2(K))
      BN(KK)=B(KK)
      BO(K)=(2.0*B(K)/BN(20)-1.0)*0.05*BN(20)**0.75
      WRITE(L7,102)K,ZDYS,ZPTOX,ZX,ZXPOS1,ZXPOS2,B1(K),B2(K),
     * S0(K),S1(K),S2(K)
      GO TO 81
   35 TX=C(92)+C(93)-AMOD(TT+C(92),C(93))
      GO TO 81
   36 SS=S(KK)*AMAX1(0.0,AMIN1((B(KK)-C(67))/C(68),1.0))
      IF(S(KK).LT.5.0) SS=S(KK)
      B(KK)=B(KK)-C(69)*SS/S(KK)
      S(KK)=S(KK)-SS
      OK(KK)=OK(KK)+SS*C(78)*(B(KK)+C(69))
      IF(-EL(K).LT.C(40).OR.AGE(KK).GT.C(94)) GO TO 37
      TX=C(93)
      GO TO 81
   37 IK(KK)=9
      CALL MEMO (KK,3,TT+C(93))
      EL(KK)=0.0
      EL(K)=0.0
      KY(K)=-KY(K)
      IK(K)=1
      IDRAFT=1
      IF(K.EQ.1) GO TO 38
      CALL MEMO (4,3,TT+TI(1)-TK(K))
      TX=1000.0
      GO TO 81
   38 TX=TI(1)-TK(K)-C(96)
      GO TO 81
   45 SS=S(K)*(AMAX1(0.0,AMIN1((B(K)-C(67))/C(68),1.0)))
      IF(S(K).LT.5.0.OR.AGE(K).GT.C(95)) SS=S(K)
```

```
      B(K)=B(K)-C(69)*SS/S(K)
      S(K)=S(K)-SS
      OK(K)=OK(K)+SS*C(78)*(B(K)+C(69))
      TX=C(93)
      GO TO 81
50    S(K)=C(81)*(1.0+C(83))-S(1)-S(2)-S(3)
      B(K)=C(84)
      BN(K)=B(K)
      W(K)=C(85)
      AGE(K)=C(86)
      OSK=S(K)*VK(IK(K))*AMAX1(0.0,AMIN1(AGE(K)/180.0,1.0))
      OK(K)=OK(K)-OSK/(1.0+100.0*EXP(-10.0*(2.0*B(K)-BN(K))/BN(K)))
      TX=1000.0
      IF(AGE(K).LT.365.0) TX=365.0-AGE(K)
      GO TO 81
61    IF(K.EQ.1) GO TO 71
      TX=1000.0
      GO TO 81
71    DO 74 K1=1,NK
      K2=KRE(K1)
      IF (K2.EQ.K1) GO TO 74
      IF (S(K2).GT.1.0) GO TO 73
      IK(K2)=IK(K1)
      TK(K2)=TK(K1)
      T(K2,3)=T(K1,3)
73    T(K1,3)=1000.0
      S(K2)=S(K2)+S(K1)+.001
      SS=S(K1)/S(K2)
      B(K2)=B(K2)+SS*(B(K1)-B(K2))
      BN(K2)=BN(K2)+SS*(BN(K1)-BN(K2))
      DK(K2)=DK(K2)+SS*(DK(K1)-DK(K2))
      AGE(K2)=AGE(K2)+SS*(AGE(K1)-AGE(K2))
      W(K2)=W(K2)+SS*(W(K1)-W(K2))
      S(K1)=0.0
      B(K1)=0.0
      AGE(K1)=0.0
      W(K1)=0.0
74    CONTINUE
      DO 75 K1=1,NK
75    CALL MEMO (K1,3,T(K1,3))
      OV(K)=-C(77)*S(K)
      SS=DIM(S(1)+S(2)+S(3),C(81))
      S(K)=S(K)-SS
      OSK=SS*VK(IK(K))*C(75)*AMAX1(0.0,AMIN1(AGE(K)/180.0,1.0))
      OK(K)=OK(K)+OSK/(1.0+100.0*EXP(-10.0*(2.0*B(K)-BN(K))/BN(K)))
      AGE(K)=635.0+TT
      TX=C(96)
81    IF(TX.EQ.0.0) TX=TI(IK(K))-TK(K)
      TK(K)=TK(K)+TX
      CALL MEMO(K,3,TT+TX)
91    DO 95 KX=1,NK
      IF(S(KX).GT.1.0) GO TO 93
      IF(MK(KX).EQ.0) GO TO 95
      DO 92 NX=KX,120,20
92    S(NX)=0.0
      CALL MEMO (KX,3,1000.0)
      GO TO 94
93    IF(MK(KX).NE.0) GO TO 95
94    IDRAFT=1
95    CONTINUE
      IF(IDRAFT.EQ.1) CALL MEMO(5,1,T(1,1)-0.2)
      IF(TT.LT.TLU(7)) RETURN
      WRITE(L7,101) TT,K,IK(K),TK(K),AGE(K),(S(L),L=1,8),(B(L),L=1,8)
      RETURN
      END
```

10 Subroutine DRAFT

Subroutine DRAFT carries out two types of operation needed in grazing management: the formation of mobs from the various stock classes and the allocation of fodder units to mobs. The variables in the subroutine are defined in Table 13. The routine may be called from the main program for a variety of reasons:

a. to evaluate the feed situation and stock distribution at regular intervals, including those determined by rotational grazing,

b. to deal with a change in status of one or more fodder units, e.g. paddocks being shut up for hay, and

c. to meet a change in status of a stock class resulting in a change in priority value, PK(I), and hence in feed requirements.

In the first two cases, the subroutine merely re-allocates fodder units to the existing mobs, but in the third case, the mobs must first be reconstituted so that stock classes with similar priorities can be brought together. This latter operation, which forms the first part of the subroutine (Fig. 20), is carried out following a call from REPRO to subroutine MEMO. In this case the parameter IX will be greater than zero (line 23).

The priority value of each stock class is determined by not only its status but its body weight relative to normal body weight. If actual body weight is less than normal body weight, then the priority value is increased by the difference in weight, expressed as a proportion of normal body weight, multiplied by the management parameter C(5). If body weight is greater than normal body weight, the priority value is reduced in a similar way, except that the factor used is management parameter C(4). For example, if normal body weight is 50 kg and actual weight is only 40 kg, the priority value is increased by 0.2 times C(5).

With mob number M initially set at zero, and the parameter PKI at some very large value (line 28), each class K is considered in order of priority value. This is done by following the

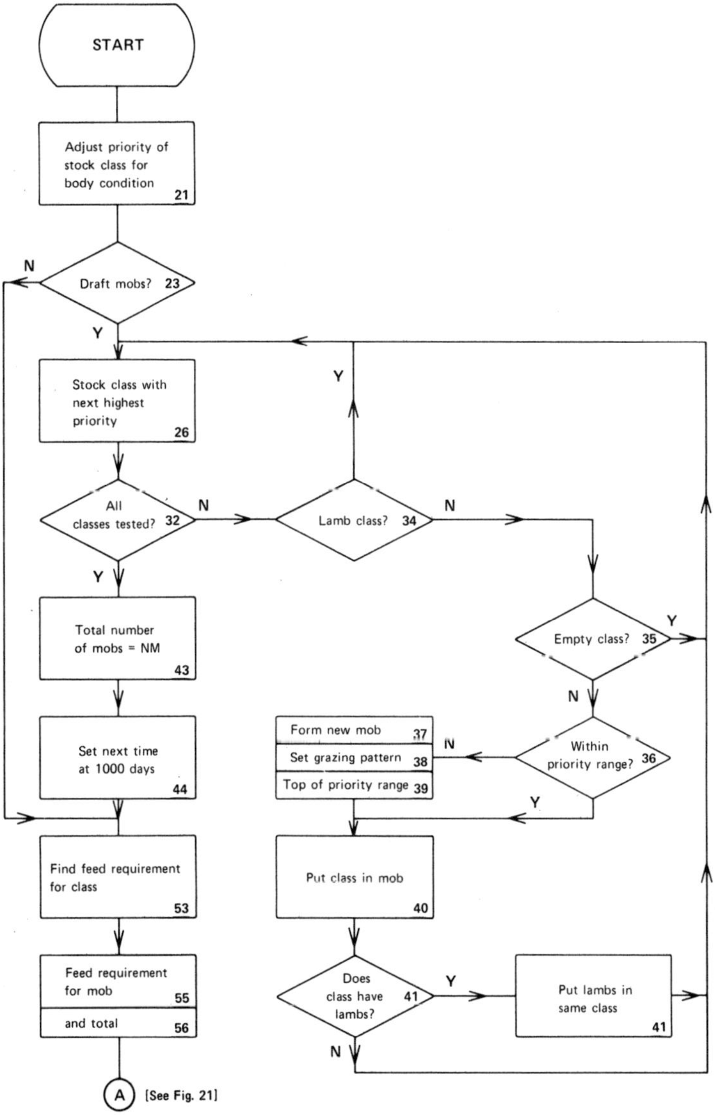

Fig. 20. Flow chart for subroutine DRAFT (part 1); the number in the corner of each box indicates the appropriate line in the listing.

Table 13 Definition of variables in subroutine DRAFT

Variable name	Definition
A(J)	Area of fodder unit J (ha)
DJP	Total feed supply of available fodder units (kg)
DKS	Feed requirement of stock class (kg)
DM(3,M)	Total potential intake for all sheep in mob M (kg)
DM(4,M)	Feed requirements for all sheep in mob M (kg)
DSP	Feed requirements for all sheep (kg)
FI(I)	Grazing pattern (rotational or set-stocking) for class K with status I = IK(K)
FM(M)	Grazing pattern for mob M
FR	Proportional feed requirement of mob
IK(K)	Status of class K
IX	Flag indicating whether mobs are to be reconstituted or not
KY(K)	Class holding the lambs from ewes in class K
M	No. of mob
MJ(J)	Mob on fodder unit J
MK(K)	Mob to which stock class K belongs
NJP	No. of fodder units available for feed consumption
NM	Total number of mobs
NT(N,NN)	See MEMO
PJ(I)	Priority value of fodder unit J with status I = IJ(J)
PK(I)	Priority value of stock class K with status I = IK(K)
PKK(K)	Priority value of stock class K, as modified by body weight
PKI	Highest priority value in mob under consideration
RD(6,1)	Relative intake of hay
RD(6,J)	Relative intake on unit J
RDT	Total relative intake of fodder units available to mob
S(K)	No. of sheep in class K

sequence of numbers in row 5 of the NT array, starting with K = NT(20,5) and ending when NT(K,5) = 20 (Fig. 20). If the class consists of unweaned lambs, IK(K) = 8, it is not treated separately (line 34), but is put into the same mob as the ewes to which they belong (line 41). Empty classes are ignored (line 35).

The management parameter C(7) determines the range of priority values which are to be permitted within the one mob (line 36). Whenever this range is exceeded, a new mob is formed by increasing M by one, and making PKI the new upper priority value limit (line 39). For instance, if the PKK(K) values for the first 4 classes were (1) 0.9, (2) 0.8, (3) 0.6 and (4) 0.4, with C(7) = 0.25, the first 2 classes would go into mob 1, and the third and fourth into mob 2. The grazing management pattern FM(M) is made the same as that for the first class to enter the mob, FI(I) (line 38). In the example just given, whether mob 1 would be set-stocked or rotationally grazed would depend on the pattern stipulated for the first class, while mob 2 would be grazed according to the pattern set down for class 3. The total number of mobs is NM. The call to subroutine MEMO (line 44) prevents redrafting until requested by some other part of the program.

The allocation of fodder units to mobs starts with the calculation of the total potential intake for each class, DKS (line 53), and for each mob, DM(3,M) (line 54). The feed requirement, defined as the product of potential intakes and priority values, is summed for each mob, DM(4,M) (line 55) and for all animals, DSP (line 56).

The relative intake of hay, RD(6,1) is calculated from the same function of digestibility as was used for herbage in subroutine FODDER, assuming availability to be non-limiting.

The subroutine next decides which fodder units are to be made available to the stock, according to the current feed position in relation to demand (Fig. 21). Taking the fodder units in increasing order of priority, according to row 4 of the NT array, the priority value is multiplied by the feed supply, DJP, which is the cumulative sum of products of area and relative intake as each fodder unit is considered (line 69).

If the product of priority and supply is greater than the total feed requirements, and provided that at least one paddock is available for each mob (line 67), the fodder unit is made unavailable for consumption by putting RD(6,J) = 0 (line 74) and removing any mob already present (line 76).

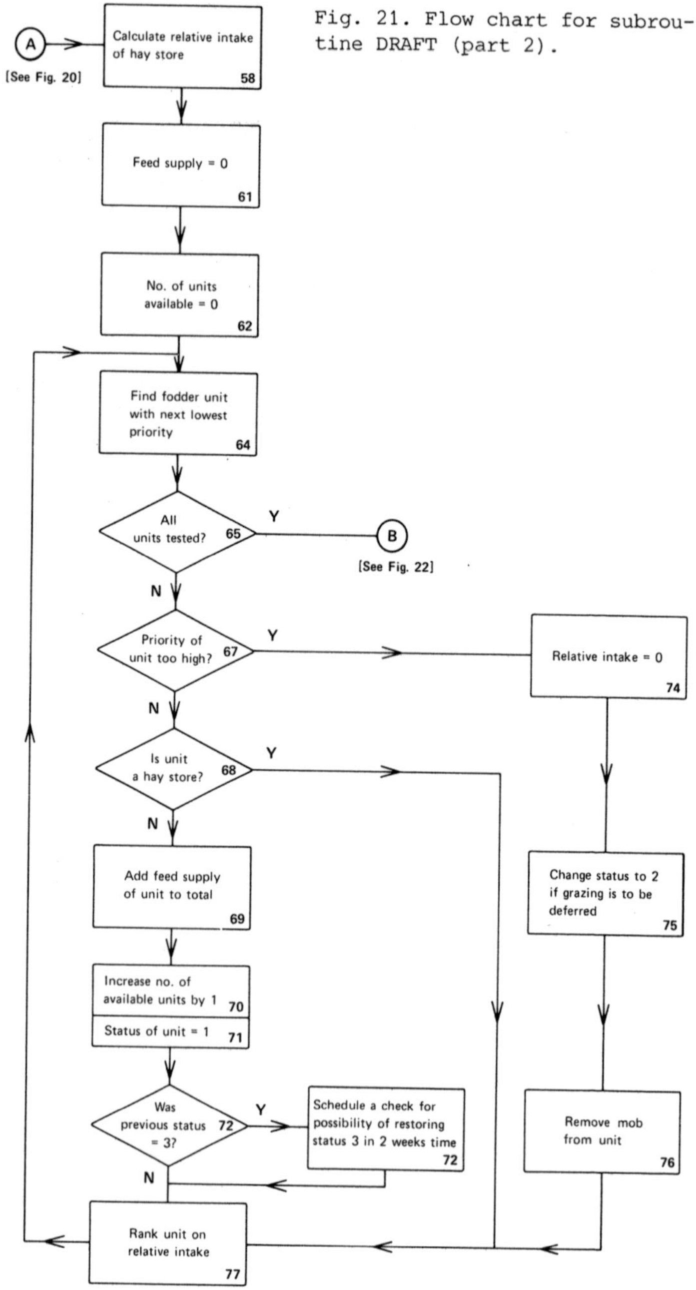

Fig. 21. Flow chart for subroutine DRAFT (part 2).

It follows that all succeeding fodder units will also be disqualified, since they have still higher PJ values. If the ratio of the product of priority and feed supply to feed requirement exceeds a factor, C(18), the paddock is closed for deferred grazing (line 75).
The status of all available paddocks is set at 1 (line 71).
If the paddock had been saved for hay up to that time (I = 3), a call is scheduled in another 14 days time (line 72), to check whether the feed situation has improved enough by then to enable the paddock to be shut up for hay again.
Each fodder unit is then arranged in descending order of relative feed intake, RD(6,J), in row 6 of the NT array by calling subroutine MEMO (line 77).

In the next stage (Fig. 22), a complete redistribution of all available fodder units is made possible only if the mobs have just been redrafted (i.e. when IX = 1) (line 81).
The feed supply on each unit (area x relative intake) is expressed as a proportion of the total, DJP, and stored as RD(1,J) (line 85). Subroutine MEMO is also called to update scheduling times and rank priorities.

Finally, fodder units are allocated to each mob, (Fig. 22), starting with the mob of highest priority and fodder units of highest relative intake, RD(6,J), and attempting as far as practicable to match the proportional needs of each mob to the proportion of the total supply offered by each of the paddocks. When a rotational grazing system is chosen, the mob will have access to only a proportion (1/FM(M)) of the allotted paddocks at one time. It follows that for set-stocking, FM(M) = 1.
As shown in line 83, if redrafting of mobs has not occurred, only those paddocks under rotational grazing will be redistributed.

The requirements of each mob as a fraction of the total are given by FR = DM(4,M)/(FM(M)*DSP) (line 89). The allocation of units to each mob ceases either when its proportional requirements are exceeded, i.e. when FR is less than RDT, or when there are only enough units left to ensure that at least one is reserved for each remaining mob (line 96). Furthermore, those units which are already set-stocked by any mob cannot be allocated to any other mob (line 95). A unit is assigned to a mob by putting MJ(J) = M (line 97), reducing the number of available units NJP by one, and adding the feed supply to the total of the mob, RDT (line 99).
The subroutine ends by scheduling another reallocation of units in C(6) day's time (line 102).

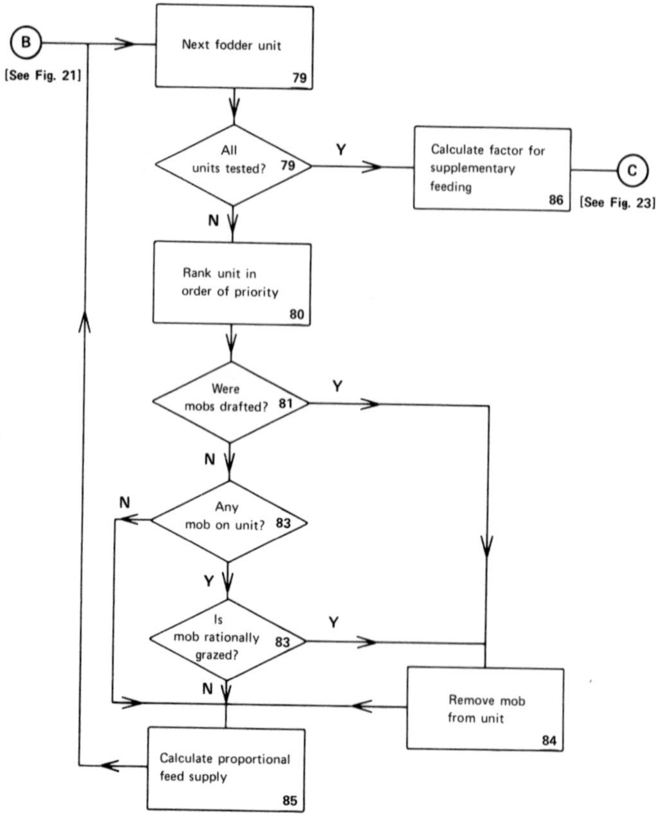

Fig. 22. Flow chart for subroutine DRAFT (part 3).

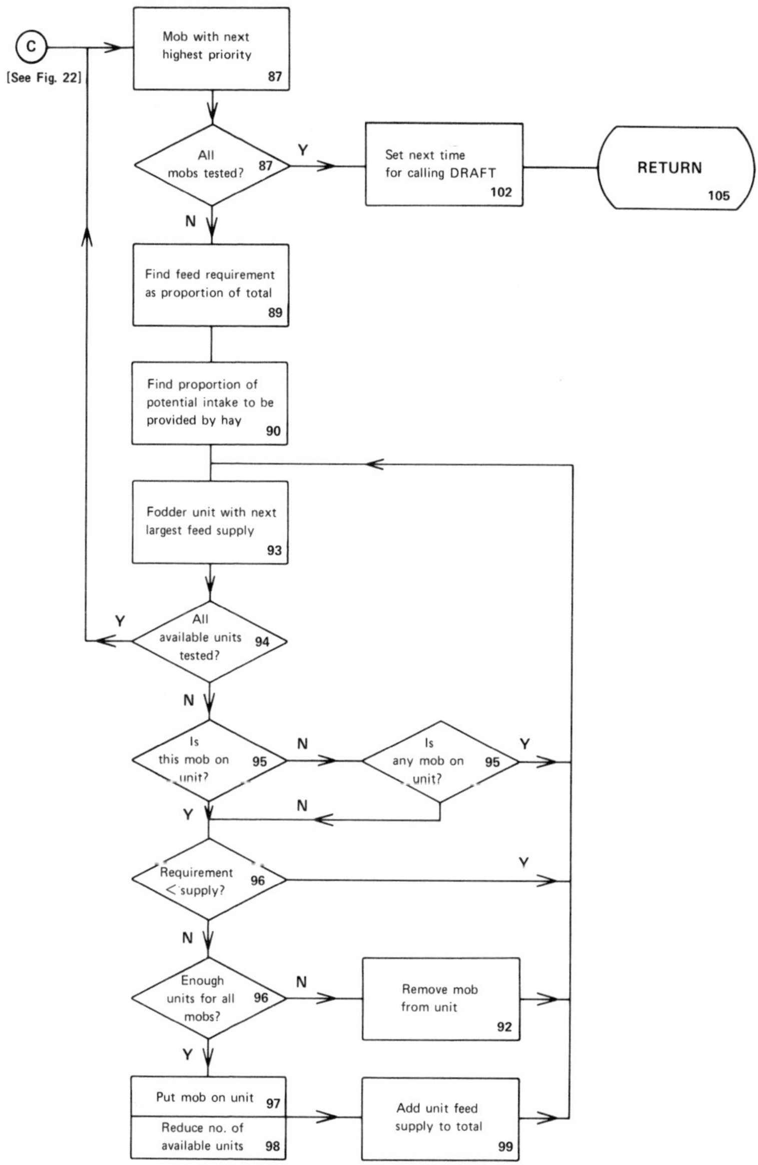

Fig. 23. Flow chart for subroutine DRAFT (part 4).

Program subroutine DRAFT

```
      SUBROUTINE DRAFT (IX)
      COMMON/A/RD(6,20),DM(4,10),FM(10),SM(10),QM(10),TT,TYR,DFT
      COMMON/B/T(20,6),F(6,20),FL(5,3),Q(5),CFG(27,3),FG(3),TLU(10)
      COMMON/C/C(100),A(20),S(20),B(20),BN(20),AGE(20),W(20),
     * DI(20),EL(20),PJ(20),PK(20),VJ(20),VK(20),FI(20),TI(20),TK(20)
      COMMON/D/D(20),DK(20),E(20),BG(20),MJ(20),MK(20),NT(20,6)
      COMMON/Y/S0(2),S1(2),S2(2),B1(2),B2(2),B0(2),TMID(2)
      COMMON/O/OJ(20),OK(20),OW(20),OV(20),XOP(40),COP(20),DOP(20),
     * IOP(20),NOP,NRUN,NDO,IYR,IO,IRUN,IDO,IREV,IMP,OO,OOO,OEI,OEIT
      COMMON/M/NFG(20),IJ(20),IK(20),KRE(20),KY(20)
      COMMON/N/NJ,NK,NL,NM,NP,NDP,NYR,LU,LFG,L1,L2,L3,L4,L5,L6,L7,L8
      DIMENSION PKK(20)
  101 FORMAT(F5.0,2X,6I2,2X,6(F6.0,I2),2X,8(F5.2,I2))
      DO 2 K=1,NK
      IF(S(K).GT.1.0) GO TO 1
      PKK(K)=0.0
      GO TO 2
    1 I=IK(K)
      BF=C(5)
      IF(BN(K).GT.B(K)) BF=C(4)
      PKK(K)=PK(I)*(1.0+BF*(BN(K)-B(K))/(BN(K)+.001))
    2 CONTINUE
      IF(IX.EQ.0)  GO TO 11
      DO 3 K=1,NK
      MK(K)=0
    3 CALL MEMO (K,5,-PKK(K))
      M=0
      PKI=999.9
      K=20
    4 K=NT(K,5)
      KK=KY(K)
      IF (K.EQ.20) GO TO 6
      I=IK(K)
      IF (I.EQ.8) GO TO 4
      IF (S(K).LT.1.0) GO TO 4
      IF(PKI-PKK(K).LT.C(7)) GO TO 5
      M=M+1
      FM(M)=FI(I)
      PKI=PKK(K)
    5 MK(K)=M
      IF(KK.GT.0) MK(KK)=M
      GO TO 4
    6 NM=M
      CALL MEMO(5,1,1000.0)
   11 IF(NM.EQ.0.0) GO TO 51
      DO 12 M=1,NM
      DM(3,M)=0.0
   12 DM(4,M)=0.0
      DSP=0.001
      DO 20 K=1,NK
      M=MK(K)
      IF(M.EQ.0) GO TO 20
      DKS=DK(K)*S(K)
      DM(3,M)=DM(3,M)+DKS
      DM(4,M)=DM(4,M)+DKS*PKK(K)
      DSP=DSP+DKS*PKK(K)
   20 CONTINUE
      RD(6,1)=C(17)*(1.0-2.0*(0.8-C(19)))
      DO 21 J=1,NJ
   21 CALL MEMO (J,4,PJ(IJ(J))+.0001*F(6,J))
      DJP=0.001
      NJP=0
      J=20
   22 J=NT(J,4)
      IF(J.EQ.20) GO TO 31
```

```
      I=IJ(J)
      IF(PJ(I)*DJP.GT.DSP.AND.NJP.GE.NM) GO TO 25
      IF(I.EQ.4) GO TO 30
      DJP=DJP+RD(6,J)*A(J)
      NJP=NJP+1
      IJ(J)=1
      IF(I.EQ.3) CALL MEMO(J,2,AMIN1(T(J,2),TT+13.9))
      GO TO 30
   25 RD(6,J)=0.0
      IF(I.EQ.1.AND.C(18)*PJ(2)*DJP.GT.DSP) IJ(J)=2
      MJ(J)=0
   30 CALL MEMO (J,6,-RD(6,J))
      GO TO 22
   31 DO 40 J=2,NJ
      CALL MEMO(J,4,PJ(IJ(J))+.0001*F(6,J))
      IF(IX.GT.0) GO TO 39
      M=MJ(J)
      IF(M.EQ.0.OR.FM(M).EQ.1) GO TO 40
   39 MJ(J)=0
   40 RD(1,J)=RD(6,J)*A(J)/DJP
      FSUP=RD(6,1)*DIM(DSP,DFT)/DSP
      DO 50 M=1,NM
      RDT=0.0
      FR=DM(4,M)/(FM(M)*DSP)
      DM(4,M)=FSUP*DM(4,M)/(DM(3,M)+0.001)
      J=20
   42 MJ(J)=0
   44 J=NT(J,6)
      IF (RD(6,J).LT.0.01) GO TO 50
      IF(MJ(J).NE.M.AND.MJ(J).NE.0) GO TO 44
      IF (FR.LT.RDT.OR.NJP.LE.NM-M) GO TO 42
      MJ(J)=M
      NJP=NJP-1
      RDT=RDT+RD(1,J)
      GO TO 44
   50 CONTINUE
   51 CALL MEMO (4,1,C(6)+T(1,1)-0.01)
      IF (TLU(8).GT.0.0) WRITE(L8,101)TT,(IJ(J),J=1,6),
     * (F(6,J),MJ(J),J=1,6),(PKK(K),MK(K),K=1,8)
  100 RETURN
      END
```

11 Subroutine OPTI

This subroutine evaluates the objective function and directs the course of optimization runs to obtain the maximum value for this function. The subroutine is called at the start of each run, with IRUN = 0, and at the end of each year of simulation. Variables used in the subroutine are defined in Table 14.
The objective function is made up of four components: value of fodder OJ(J), stock OK(K), and wool OW(K), and variable costs per animal OV(K) (Fig. 24).

The subroutine firstly prints out the net value of income derived and expenses incurred for each fodder unit and stock class over the course of the year (line 24). Purchases of hay and costs of haymaking are debited against the appropriate fodder units in the OJ array, and sales and purchases of stock make up the OK values. These values are summed on a per hectare basis and stored as OEI (lines 29 and 38).
The OJ and OK arrays are then used to store the estimated values of fodder units and stock classes on the property at the end of the year. The value of fodder on each paddock is assessed according to the weight of dry matter and the digestibility of each class (line 33). The value ($\$ha^{-1}$) over the whole farm is obtained as the weighted mean according to the area of individual paddocks, and added to OJ(20) (line 34). The value per ha for each stock class is a function of total numbers, total farm area, and body weight and condition (lines 40 and 41). Wool is assumed for convenience to be removed at the end of each year (lines 44 and 56).

The cumulative total of OEI, wool sales and variable costs, less the value per hectare of fodder and stock at the start of the run, is stored as OEIT (lines 46 and 47). The objective function OO is therefore obtained each year as the increase in value of feed and livestock and the excess of income over expenses during the period to date. A call to subroutine MEMO ensures that the objective function is calculated again in a year's time (line 57), and no further action is taken until the nominated number of years for simulation, NYR, is completed (line 61).

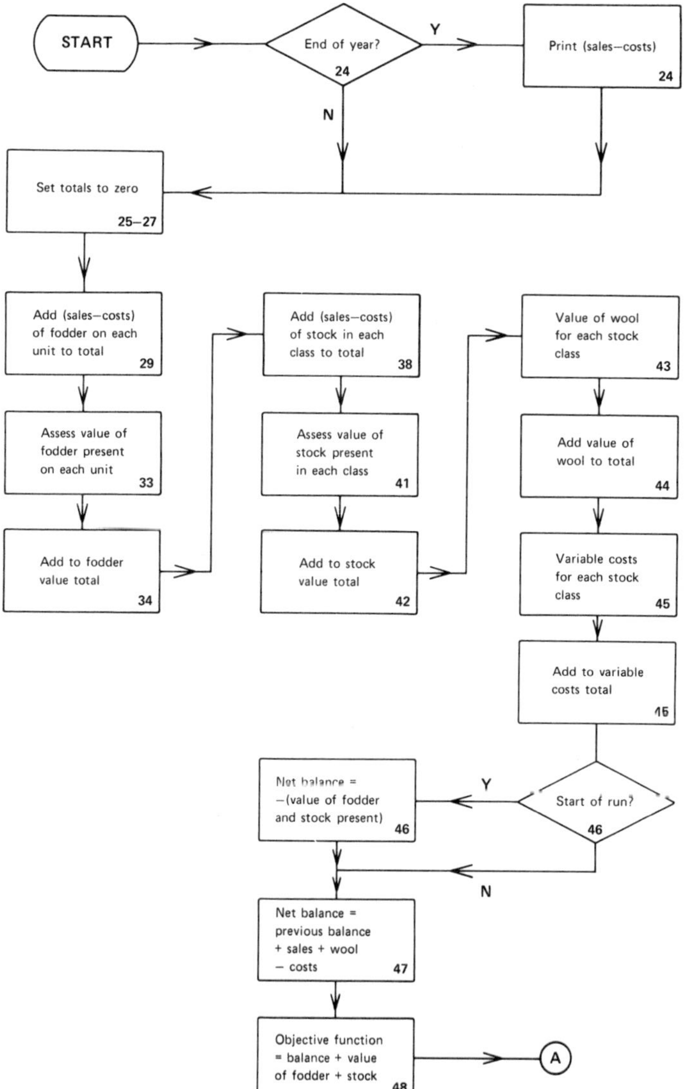

Fig. 24. Flow chart for subroutine OPTI (part 1); the number in the corner of each box indicates the appropriate line in the listing.

Table 14 Definition of variables in subroutine OPTI

Variable name	Definition
COP(IO)	Best value so far for parameter IO
DOP(IO)	Step size in varying parameter IO
IDO	No. of reductions in step size so far
IMP	Flag denoting improvement in objective function
IO	Index in IOP array of parameter currently being optimized
IOP(IO)	Index in C array of parameter IO
IREV	Flag denoting whether step has been made in reverse direction
IRUN	No. of current optimization run
IYR	No. of current year
NDO	Total no. of step size reductions to be made
NOP	Total no. of optimization parameters
NRUN	Total no. of optimization runs
NYR	Total no. of years in simulation run
OEI	Net returns from sale of stock after purchase or making of hay ($\$ha^{-1}$)
OEIT	All returns, less variable costs, plus increase in value of fodder and stock ($\$ha^{-1}$)
OJ(J)	Total value of fodder unit J ($\$ha^{-1}$)
OK(K)	Total value of stock class K ($\$ha^{-1}$)
OO	Objective function for current run ($\$ha^{-1}$)
OOO	Highest objective function value so far ($\$ha^{-1}$)
OV(K)	Variable costs for stock class K ($\$ha^{-1}$)
OW(K)	Total value of wool for stock class K ($\$ha^{-1}$)
VJ(I)	Value of fodder on unit of status I = IJ(J) ($\$kg^{-1}$)
VK(I)	Value of stock of class I = XIK(K) ($\$kg^{-1}$)
XOI	Coded value of parameter combinations
XOP (IRUN)	Code value of parameter combination for run IRUN

The parameters chosen for optimization, NOP in number, are specified in the IOP array by regarding them as members of the C array in the COMMON/C/ block. For instance, if A(20) is the first parameter selected, this location corresponds to C(120), and hence the value of IOP(1) read in is 120. The initial value of A(20) is then stored as COP(1) in the main program.

The general method for optimization operates in the following manner. The initial parameter values are set somewhere near the middle of the anticipated range; the model is run for a number of years under a standard set of seasonal pasture conditions, and the objective function is calculated at the end of this time. The first parameter under test is then increased by a pre-determined step size to a value somewhere near the top of the anticipated range, the run is repeated with this new value, and the objective function is recalculated. If the objective function is higher, the new parameter value is retained; if lower, a step is made in the reverse direction. If neither move produces any improvement, the original value is restored. The routine tests all the parameters in turn in this way, repeating the cycle as long as progress continues to be made. When all further moves fail, the step size is reduced, and the area around the optimum examined more closely; this process may be repeated as long as it is considered that further refinement is justified.

However, the program steps, shown in Fig. 25, follow a somewhat different order. At the end of any run in which the current value of the IOth parameter is being tested against a previous value, the resulting objective function, OO, is compared with the previous best value for this parameter, OOO (line 62). If there is an improvement, OO becomes the best value (line 63), and the flag IMP is set at 1. If there is no improvement, the COP value for the IOth parameter is reset at its previous value (line 66). If IREV = 0, indicating that a step in the reverse direction has not yet been undertaken (line 67), the sign of the step size DOP(IO) is changed (line 68), and the flag IREV set at 1 (line 69).

When all the parameters have been tested (line 71) and there is no improvement in the objective function (line 72), the step size is halved (line 74), and the counter IDO is incremented. A new cycle starts at line 77, taking each parameter in turn (line 79), and incrementing the best value so far by the appropriate step value (line 81).
To prevent the repetition of runs having the same combination of optimization parameter values, a record of the set for each run is made in the form of a single coded number XOI,

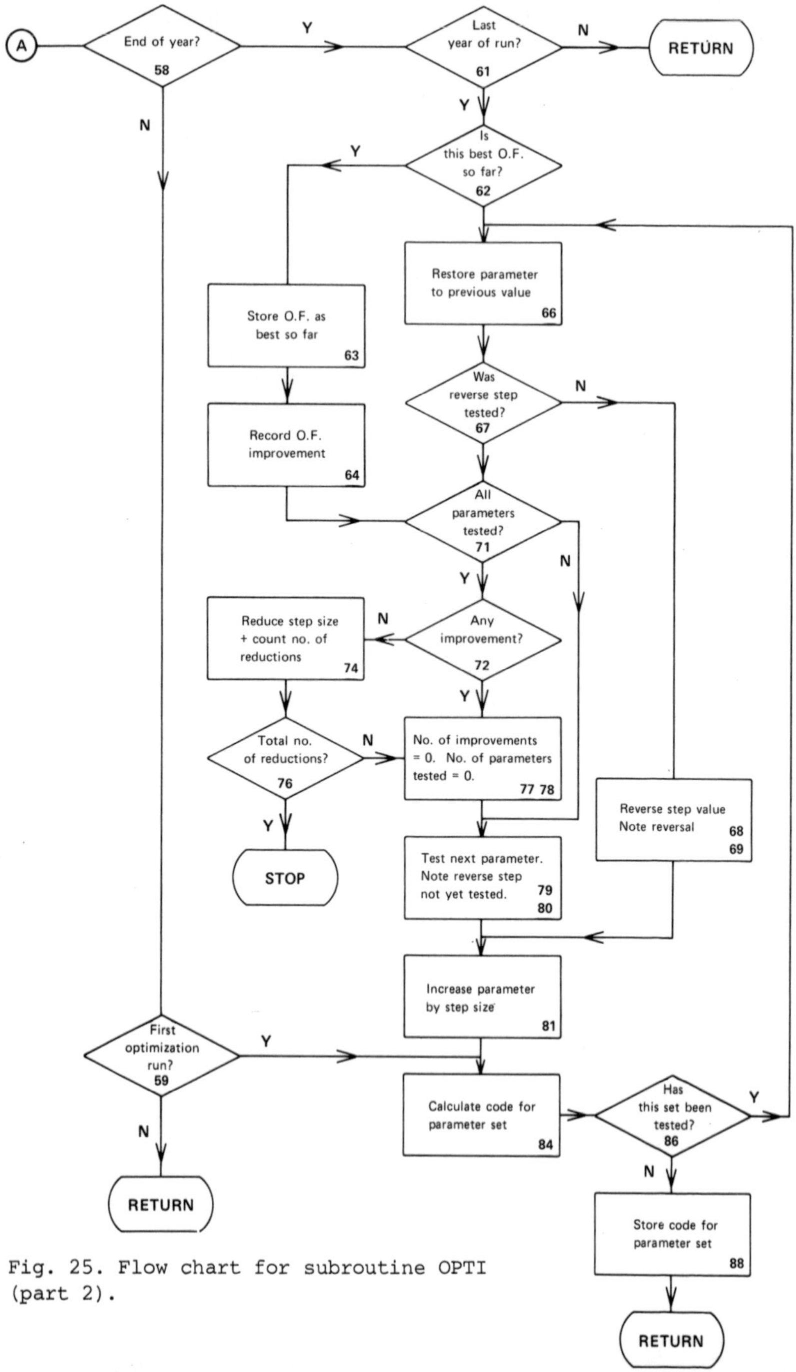

Fig. 25. Flow chart for subroutine OPTI (part 2).

which is computed from the parameter values in such a way that it is most unlikely to be reproduced by any other combination of variables (line 84). This XOI value is compared with the values obtained from previous runs and stored in the XOP array (line 88). If the figures match, the routine tries the next step in the series until all possibilities have been exhausted. If a new combination is found, another run is started. The procedure continues until either the maximum number of runs, NRUN, or the maximum number of step size reductions, NDO, has been reached (line 76).
The validity of the method rests on the assumption that the response function for each parameter is unimodal, and the results obtained from one set of runs should be compared with those arrived at from different starting points and through different step sizes.

Program subroutine OPTI

```
 1            SUBROUTINE OPTI
 2            COMMON/A/RD(6,20),DM(4,10),FM(10),SM(10),QM(10),TT,TYR,DFT
 3            COMMON/B/T(20,6),F(6,20),FL(5,3),Q(5),CFG(27,3),FG(3),TLU(10)
 4            COMMON/C/C(100),A(20),S(20),B(20),BN(20),AGE(20),W(20),
 5        *   DI(20),EL(20),PJ(20),PK(20),VJ(20),VK(20),FI(20),TI(20),TK(20)
 6            COMMON/D/D(20),DK(20),E(20),BG(20),MJ(20),MK(20),NT(20,6)
 7            COMMON/Y/S0(2),S1(2),S2(2),B1(2),B2(2),B0(2),TMID(2)
 8            COMMON/O/OJ(20),OK(20),OW(20),OV(20),XOP(40),COP(20),DOP(20),
 9        *   IOP(20),NOP,NRUN,NDO,IYR,IO,IRUN,IDO,IREV,IMP,OO,OOO,OEI,OEIT
10            COMMON/M/NFG(20),IJ(20),IK(20),KRE(20),KY(20)
11            COMMON/N/NJ,NK,NL,NM,NP,NDP,NYR,LU,LFG,L1,L2,L3,L4,L5,L6,L7,L8
12     101    FORMAT(/' INCOME-EXPENSES FOR FODDER UNITS'/1X,19F6.0,F8.1//
13        *   ' INCOME-EXPENSES FOR STOCK CLASSES'/1X,19F6.0,F8.1//)
14     102    FORMAT(' VALUES OF OPTIMIZATION PARAMETERS'//1X,20F6.1/)
15     103    FORMAT(' TOTAL FEED/HA. ON EACH UNIT'/1X,19F6.0,F8.1//
16        *   ' STOCK NUMBER IN EACH CLASS'/1X,19F6.0,F8.1//
17        *   ' STOCK BODYWEIGHTS'/1X,19F6.1,F8.1//
18        *   ' WOOL WEIGHTS'/1X,19F6.2,F8.1//)
19     104    FORMAT(' TOTAL VALUE OF FEED ON FODDER UNITS'/1X,19F6.0,F8.1//
20        *   ' TOTAL VALUE OF STOCK IN EACH CLASS'/1X,19F6.0,F8.1//
21        *   ' VALUE OF WOOL FOR EACH STOCK CLASS'/1X,19F6.0,F8.1//
22        *   ' VARIABLE COSTS FOR EACH STOCK CLASS'/1X,19F6.0,F8.1//
23        *   ' OBJ.F FOR THIS RUN',F10.2,5X,'PREVIOUS BEST OBJ.F.'F10.2/)
24            IF(TT.GT.1.0.AND.AMOD(TT,TLU(1)).LT.0.9) WRITE(L1,101)OJ,OK
25            OEI=0.0
26            OJ(20)=0.0
27            OK(20)=0.0
28            DO 4 J=1,NJ
29            OEI=OEI+OJ(J)/A(20)
30            OJ(J)=0.0
31            I=IJ(J)
32            DO 3 L=1,NL
33       3    OJ(J)=OJ(J)+VJ(I)*F(L,J)*A(J)*(Q(L)-C(76))/(0.60-C(76))
34       4    OJ(20)=OJ(20)+OJ(J)/A(20)
35            S(20)=0.0
36            DO 5 K=1,NK
37            S(20)=S(20)+S(K)
38            OEI=OEI+OK(K)/A(20)
39            I=IK(K)
40            OK(K)=S(K)*VK(I)*AMAX1(0.0,AMIN1(AGE(K)/180.0,1.0))
41            OK(K)=OK(K)/(1.0+100.0*EXP(-10.0*(2.0*B(K)-BN(K))/(BN(K)+.001)))
42            OK(20)=OK(20)+OK(K)/A(20)
```

```
      OW(K)=C(74)*W(K)*S(K)
      OW(20)=OW(20)+OW(K)/A(20)
    5 OV(20)=OV(20)+OV(K)/A(20)
      IF(TT.LT.365.0) OEIT=-OJ(20)-OK(20)
      OEIT=OEIT+OEI+OW(20)+OV(20)
      OO=OEIT+OJ(20)+OK(20)
      IF(AMOD(TLU(1),365.0).GT.0.9) GO TO 6
      WRITE(L1,103) (F(6,J),J=1,20),S,B,W
      WRITE(L1,104) OJ,OK,OW,OV,OO,OOO
      WRITE(L1,102) (COP(I),I=1,NOP)
    6 DO 7 I=1,80
    7 OJ(I)=0.0
      DO 8 K=1,NK
    8 W(K)=0.0
      CALL MEMO (6,1,TT+365.0)
      IF(TT.GT.365.0) GO TO 9
      IF(IRUN.EQ.1) GO TO 24
      RETURN
    9 IF(IYR.NE.NYR) RETURN
      IF (OO.LE.OOO)  GO TO 12
      OOO=OO
      IMP=1
      GO TO 15
   12 COP(IO)=COP(IO)-DOP(IO)
      IF (IREV.EQ.1) GO TO 15
      DOP(IO)=-DOP(IO)
      IREV=1
      GO TO 22
   15 IF (IO.LT.NOP) GO TO 20
      IF (IMP.EQ.1) GO TO 18
      DO 16 I=1,NOP
   16 DOP(I)=DOP(I)*0.5
      IDO=IDO+1
      IF(IDO.GT.NDO) STOP
   18 IMP=0
      IO=0
   20 IO=IO+1
      IREV=0
   22 COP(IO)=COP(IO)+DOP(IO)
   24 XOI=0.0
      DO 25 I=1,NOP
   25 XOI=XOI+(I+0.12345)*(COP(I)+3.14159)
      DO 30 IX=1,IRUN
      IF (ABS(XOI-XOP(IX)).LT.0.0001) GO TO 12
   30 CONTINUE
      XOP(IRUN)=XOI
      RETURN
      END
```

12 Example

Setting up the model

Before attempting to use the model, the operator should carefully consider what categories of animals he wishes to include. It is suggested that a flow chart, along the lines of Fig. 2. should be drawn up to indicate the points at which transitions from one stock class or status to another are required.
The necessary adjustments can then be made in subroutine REPRO.
Assuming however, that the structure shown in Fig. 2 is suitable, the operator can proceed to enter information in the general data file LU. Two broad categories of data are needed:
a. lines 1-20: initial values for status, for the scheduling of events and for other attributes which define the system being modelled and are not normally subject to optimization, and
b. lines 21-44: general constants and stock and management parameters, many of which can be the subject of optimization.
The function of each line of the general input data is shown below and the values used in this example, as entered in the data file LU, are shown in Tables 15 and 16.

a. *Line*
1. Specifies the input file LFG, containing the pasture growth parameters.
2. Specifies the output files L1 to L8. More files may be created if desired to obtain separate lists of information from various parts of the program.
3. TLU(N) specifies the intervals (days) at which data are printed on files L1-L8.
4. NRUN is the total number of optimization runs. If optimization is not required, this value is set at 1.
5. NFG (IYR) specifies the sequence in which the seasonal growth patterns in the LFG file are to be read into the CFG array.
6. IJ(J) holds the initial status values of all the fodder units. In this example, unit 1 has status 4 (hay store) while the remainder have status 1 (paddocks available for grazing).
7. IK(K) is the status value of each stock class. These values

Table 15 Example of data on file LU: lines 1-44.

```
 LFG
  12

  L1  L2  L3  L4  L5  L6  L7  L8
  13  14  15  16  17  18  19  20

TLU=TIMES FOR OUTPUT ON LOGICAL UNITS
 365   7   7   7   7   7   7   7

NRUN
   1

NFG=SEQUENCE OF SEASONAL GROWTH PATTERNS
   2   2   4   3   2   2   1   4

IJ=STATUS OF FODDER UNITS
   4   1   1   1   1   1   1   1   1   1   1   1   1   1   1   1   1   1   1   1

IK=STATUS OF STOCK CLASSES
   1  11   0   0   0   9   0   0

KRE=NEW CLASSES AFTER REDISTRIBUTION
   1   1   1   2   5   6   6   6

KY=CLASS TO WHICH YOUNG OF EACH CLASS BELONGS
  -7  -8

  NK  NL  NM  NP NDP NYR
   8   5   1  26  14   1

T(I,1)=TIMES FOR GENERAL EVENTS
 1.4100010001000 1.5  .5

T(J,2)=TIMES FOR FODDER UNIT EVENTS
1000100010001000 220 2201000100010001000100010001000100010001000100010001000

T(K,3)=TIMES FOR STOCK CLASS EVENTS
  30  50100010001000  1410001000

F(1,J)=FEED AVAILABLE IN DIGESTIBILITY CLASS 1 ON EACH UNIT
   01200120012001200120012001200120012001200120012001200120012001200120012001200

F(2,J)=FEED AVAILABLE IN DIGESTIBILITY CLASS 2 ON EACH UNIT
   0 800 800 800 800 800 800 800 800 800 800 800 800 800 800 800 800 800 800 800

F(3,J)=FEED AVAILABLE IN DIGESTIBILITY CLASS 3 ON EACH UNIT
 200 500 500 500 500 500 500 500 500 500 500 500 500 500 500 500 500 500 500 500

F(4,J)=FEED AVAILABLE IN DIGESTIBILITY CLASS 4 ON EACH UNIT
   0 300 300 300 300 300 300 300 300 300 300 300 300 300 300 300 300 300 300 300

F(5,J)=FEED AVAILABLE IN DIGESTIBILITY CLASS 5 ON EACH UNIT
   0 200 200 200 200 200 200 200 200 200 200 200 200 200 200 200 200 200 200 200

F(6,J)=TOTAL FEED AVAILABLE ON EACH UNIT
 2003000300030003000300030003000300030003000300030003000300030003000300030003000

FL(L,1)=GROWTH.      FL(L,2)=SENESCENCE. FL(L,3)=DECAY RATE. Q(L)=DIGESTIBILITY
   0   0   0   0 1.0.005.015.020.025.030.995.995.995.995.995  .4  .5  .6  .7  .8

C  1   2   3   4   5   6   7   8   9  10  11  12  13  14  15  16  17  18  19  20
   0              .5  .2  14 .19   6      .1 500.004 220 350  .6  .6 1.0 1.2 .60  50

C 21  22  23  24  25  26  27  28  29  30  31  32  33  34  35  36  37  38  39  40
          .7 .08  16.257.056 2.8.026.014      .115.352.66.045 .70.017 1.4 .05  .2

C 41  42  43  44  45  46  47  48  49  50  51  52  53  54  55  56  57  58  59  60
  .51.3729.2.011 1.5.037  .1  .3  .2  .9   3 .03  10 .35   0   0 2.5 .05 .05 1.0

C 61  62  63  64  65  66  67  68  69  70  71  72  73  74  75  76  77  78  79  80
.575 1.5 .30 2.0 .05  .8  30   5   5   0             2.0 .25 .37 4.0  .3 .01
```

```
C 81   82   83   84   85   86   87   88   89   90   91   92   93   94   95   96   97   98   99 100
2500      .20   35 1.0 240 100                     17   56   14   90 240   40 150

A(J)=PROPORTIONAL AREA OF FODDER UNITS.  A(20)=TOTAL FARM AREA
   0    5    4    3    2    1    1    1    1    1    1    1    1    1    1    1    1    1    1 170

S(K)=INITIAL PROPORTIONS OF STOCK IN EACH CLASS.
   4    1    0    0    0    1    0    0

B(K)=BODYWEIGHT OF STOCK IN EACH CLASS
  55   45    0    0    0   30    0    0

BN(K)=NORMAL BODYWEIGHTS.   BN(20)=MATURE NORMAL BODYWEIGHT
  50   45    0    0    0   30    0    0                                                          50

AGE(K)=AGE OF EACH STOCK CLASS
1000 505    0    0    0 140    0  0

W(K)=WOOL WEIGHT FOR EACH CLASS
   0

DI(K)=INTAKE FACTOR FOR EACH STOCK CLASS
   1    1    1    1    1    1    1    1

EL(K)=EWE MILK PRODUCTION (NEGATIVE SIGN)
   0

PJ(I)=PRIORITIES FOR FODDER UNITS OF GIVEN STATUS
  15   40   45   60

PK(I)=PRIORITIES FOR STOCK CLASSES OF GIVEN STATUS
   .4   .8   .8   .7 1.0 1.0   .8    0 1.0   .5   .5 1.0 1.0   .7 1.0 1.0   .8

VJ(I)=VALUE PER KG.OF FEED ON UNITS OF GIVEN STATUS
 .02 .02 .02 .04

VK(I)=VALUE PER HEAD OF STOCK OF GIVEN STATUS
  20   20   20   20   20   20   20    0    5   15   20   20   20   20   20   20   20

FI(I)=GRAZING PATTERN FOR STOCK OF GIVEN STATUS
 1.5 1.5 1.5 1.5 1.5 1.0 1.5 1.5 1.5 1.0 1.5 1.5 1.5 1.5 1.5 1.0 1.5

TI(I)=TIME INTERVAL SPENT BY STOCK CLASS IN EACH STATUS
 160   21   34   74   42   34100010001000   10 160   21   34   74   42   341000

TK(K)=DAYS FROM START OF TIME INTERVAL AT WHICH NEXT STOCK EVENT OCCURS
 160 160    0    0    0    0    0    0

  S0   S0   S1   S1   S2   S2   B1   B1   B2   B2   B0   BOTMIDTMID
 1.0 1.0    0    0    0    0    0    0    0    0    0    0    0    0

IOP=LIST OF OPTIMIZATION PARAMETERS (INDICES IN C ARRAY)
  81

 NOP NDO
   1    2
 500
 DOP=INITIAL STEP SIZE FOR OPTIMIZATION PARAMETERS
 500
```

Table 16 Parameters in C array

Index	Value	Meaning
1	0	Calendar date of starting time
2	-	
3	-	
4	0.5	Priority increment for animals below normal body weight
5	0.2	Priority decrement for animals above normal body weight
6	14.0	Days between feed evaluations
7	0.19	Maximum difference between priorities within one mob
8	6	Total no. of fodder units (NJ)
9	-	
10	0.1	Proportion of herbage lost due to trampling
11	500	Minimum weight of herbage in growth equation (kg ha^{-1})
12	0.004	Exponent relating relative intake to availability
13	220	First date for closing paddocks for hay
14	350	End of haymaking period
15	0.60	Proportion of cut fodder conserved as hay
16	0.60	Minimum herbage digestibility at which to cut for hay
17	1.0	Proportion of requirement for hay actually fed
18	1.2	Priority factor in deciding to close paddock for deferred grazing
19	0.60	Mean digestibility of hay in fodder unit 1
20	50	Minimum weight of hay in fodder unit 1 (tonnes)
21	-	
22	-	This space is used by the program to hold the calculated weight of a single lamb from a ewe of normal mature body weight (kg)
23	0.7	Exponent relating normal body weight to age
24	0.08	Factor relating potential intake to mature normal body weight
25	16	Herbage dry matter equivalent of milk
26	0.257	Coefficient relating fasting body metabolism to body weight
27	0.056	Coefficient relating fasting body metabolism to energy intake
28	2.8	Coefficient relating fasting body metabolism to body weight gain
29	0.026	Factor for wool growth (related to proportion of herbage in diet)

Table 16 (continued)

Index	Value	Meaning
30	0.014	Factor for wool growth (related to proportion of herbage in diet)
31	-	
32	0.11	Factor for reducing ME in excess of maintenance
33	5.35	Factor for energy requirement of foetus
34	2.66	Exponent for energy requirement of foetus
35	0.045	Exponent for potential milk yield
36	0.7	Fraction of excess energy converted to milk
37	0.017	Factor relating intake to time after lambing
38	1.4	Factor relating intake to time after lambing
39	0.05	Factor relating intake to time after lambing
40	0.2	Minimum milk production initiating weaning
41	0.5	Efficiency of body weight loss
42	1.37	Factor relating weight gain to ME
43	29.2	Factor relating weight gain to ME
44	0.011	Exponent for wool growth according to age
45	1.5	Factor relating wool growth to ME
46	0.037	Factor relating wool growth to ME
47	0.1	Factor for stock deaths
48	0.3	Factor for stock deaths (independent of live-weight)
49	0.2	Factor for stock deaths (from undernutrition)
50	0.9	Weight of stock that die as a proportion of the weight of the survivors
51	3	Weight difference between fertile and infertile ewes (kg)
52	0.03	Factor for lamb deaths due to dystocia
53	10	Factor for lamb deaths due to dystocia
54	0.35	Factor for pregnancy toxemia deaths in ewes with twins
55	-	
56	-	
57	2.5	Factor for lamb deaths from exposure and mismothering
58	0.05	Factor for lamb deaths from exposure and mismothering
59	0.05	Factor for lamb deaths from exposure and mismothering
60	1.0	Factor for lamb deaths from exposure and mismothering
61	0.575	Factor relating probability of ovulation to body weight
62	1.5	Factor relating probability of ovulation to

Table 16 (continued)

Index	Value	Meaning
		body weight
63	0.30	Factor relating probability of ovulation to time of year
64	2.0	Exponent relating probability of fertilization to body weight
65	0.05	Factor relating probability of fertilization to time of year
66	0.8	Factor for twin ovulation
67	30	Minimum body weight for sale of lambs (kg)
68	5	Body weight above minimum at which all lambs sold (kg)
69	5	Weight difference between lambs sold and retained (kg)
70	-	
71	-	
72	-	
73	-	
74	2.0	Value of clean wool ($\$kg^{-1}$)
75	0.25	Relative value of culled ewes
76	0.37	Herbage digestibility at which value becomes zero
77	4.0	Variable costs ($ per ewe)
78	0.3	Value of lamb ($\$kg^{-1}$)
79	0.01	Cost of making hay ($\$kg^{-1}$)
80	-	
81	2500	Total no. of breeding animals (S(1) + S(2) + S(3))
82	-	No. of ewe weaners kept as proportion of C(81)
83	0.20	No. of ewe weaners bought as proportion of C(81)
84	35	Body weight of weaners bought (kg)
85	1.0	Wool weight of weaners bought (kg)
86	240	Age of weaners bought (days)
87	100	Time at which weaners bought (days)
88	-	
89	-	
90	-	
91	17	Time of ovulation cycle (days)
92	56	Days from end of lambing to start of sales
93	14	Days between sales
94	90	Maximum time after end of lambing at which lambs weaned (days)
95	240	Maximum age at which prime lambs sold (days)

Table 16 (continued)

Index	Value	Meaning
96	40	Days between class redistribution and start of flushing
97	150	Length of gestation (days)
98	-	
99	-	
100	-	

can be checked against Fig. 2. Class 1 contains the main flock of mature ewes, which are dry at the start of the year, while class 2 consists of maiden ewes, also dry. Class 6 is for the unsold lambs carried over from the previous year.

8. KRE(K) shows the stock class into which each class K will be moved in the general redistribution near the end of each year. Comparison with Fig. 2 shows that class 3 (dry ewes) and class 2 (previous season's maiden ewes) are transferred to class 1; replacement ewe weaners (class 4) become maiden ewes (class 2), and so on.

9. KY(K) indicates the class to which the young of each class belong; the negative value indicates that the ewes are neither pregnant nor lactating.

10. Contains a number of general integer constants:

NK, maximum number of stock classes present at any time.

NL, number of digestibility classes.

NM, initial number of mobs. This is set at 1 so that sorting into the appropriate number can be done automatically on the first day of simulation.

NP, number of pasture growth periods per year. It is convenient to use 26 periods of 14 days each, but these numbers can easily be altered in accordance with the LFG data file.

NDP, number of days per growth period.

NYR, number of years in each simulation run.

11-13. T(N,NN) are initial values for the times of events in each of 3 rows (where the event is not expected to occur during the running of the program, the value is set at 1000 days).

- Times for *general events*. Only three need be specified: the times for FODDER and STOCK evaluation, T(1,1), and for DRAFT, T(5,1), should occur shortly after (C(1) + 1.0), while the call to OPTI, T(6,1), should occur just less than one day earlier than T(1,1) to ensure that the year consists of 365 days. Despite the example given in Section 5, it is not necessary to provide initial values for T(2,1) to T(4,1) since these

are automatically scheduled in the program.
- Initial times for *fodder events*. The example indicates that fodder units 5 and 6 may be closed for haymaking after day 220 if feed supplies are adequate.
- Initial times for *stock events*. Those listed are: flushing of mature ewes (class 1) to start on day 30, and of maiden ewes on day 50, and inspection for the sale of class 6 weaners on day 14.
14-19. F(L,J) is the initial weight of feed dry matter (kg ha^{-1}) in each of L digestibility classes (in this case 5) on each fodder unit J. The total feed in each unit is shown as F(6,J). For fodder unit 1, the example specifies 200 tonnes of hay, all in digestibility class 3, while the amounts for all the paddocks are the same, with a total for each of 3000 kg ha^{-1}.
20. FL(L,N) and Q(L) are explained in subroutine FODDER. All new growth is assumed to be 80% digestible (i.e. FL(5,1) = 1.0), while the rate of maturation is greatest for the highest digestibility class (FL(5,2) = 0.03) and least for the class of lowest digestibility (FL(1,2) = 0.005). Material from each class is assumed to decay at a rate of 0.5% per day (FL(L,3) = 0.995). The digestibility classes range from 0.4 (Q(1)) to 0.8 (Q(5)).
21-25. C(M) contains a large number of biological constants and management parameters which are defined in Table 16.
26. A(J) is the relative area of each paddock. A(1), referring to the hay store, is set at zero. As shown in Table 15, the ratio of areas of the 5 paddocks is 5:4:3:2:1. The values shown for units 7-19 would not be used unless NJ were increased. A(20) specifies the total area of the farm.
27. S(K) are the initial proportions of stock numbers in each class. In this example, the numbers of mature ewes, maiden ewes and weaners to be sold are in the proportions 4:1:1. The total number of breeding animals is specified by C(81); hence the stocking rate can be varied without affecting flock structure.
28. B(K) is the mean body weight (kg) of stock in each class at the start of each run.
29. BN(K) is the initial normal body weight of each stock class. BN(20) is the normal body weight of a mature animal, the upper limit to BN(K).
30. AGE(K) is the initial mean age (days) of each stock class. In this example the age of the mature ewes is about 3 years and the maiden ewes about 1½ years. The age of the unsold weaners corresponds to the lambing date specified later.
31. W(K) is the weight of clean dry wool (kg) per sheep in each stock class at the start of the run. For convenience, it

Table 17 Allocation of stock classes into mobs

Day	Class changing status	New status	Stock class number															
			1		2		3		4		5		6		7		8	
	K	IK	PK[1]	MK[2]	PK	MK	PK	MK	PK	MK	PK	MK	PK	MK	PK	MK	PK	MK
1	1	1	0.4	2	0.5	2							1.0	1				
30	1	2	0.8	2	0.5	3							1.0	1				
50	2	12	0.8	2	1.0	1							1.0	1				
51	1	3	0.8	2	1.0	1							1.0	1				
71	2	13	0.8	2	1.0	1							1.0	1				
85	1	4	0.7	2	1.0	1	0.4	3					1.0	1		2		
100	4	10	0.7	2	1.0	1	0.4	3	0.5	3			1.0	1		2		
105	2	14	0.7	2	0.7	2	0.4	3	0.5	3			1.0	1		2		2
159	1	5	1.0	1	0.7	2	0.4	3	0.5	3						1		2
179	2	15	1.0	1	1.0	1	0.4	2	0.5	2						1		1
201	1	6	1.0	1	1.0	1	0.4	2	0.5	2						1		1
221	2	16	1.0	1	1.0	1	0.4	2	0.5	2						1		1
225	4	11	1.0	1	1.0	1	0.4	2	0.5	2						1		1
235	1	7	0.8	2	1.0	1	0.4	3	0.5	3						2		1
255	2	17	0.8	1	0.3	1	0.4	2	0.5	2						1		1
308	1	1	0.4	1	0.3	1	0.4	2	0.5	2					1.0	1	1.0	1
322	7	9	0.4	1	0.3	1	0.4	2	0.5	2					1.0	1	1.0	1
336	2	1	0.4	1	0.4	1	0.4	2	0.5	2					1.0	1	1.0	1
350	8	9	0.4	1	0.4	1	0.4	2	0.5	2					1.0	1	1.0	1
364	6	9	0.4	2	0.5	2							1.0	1				

1. Priority value
2. Mob number

is assumed that all sheep are shorn at the end of each year; accordingly, all the weights inserted here are zero.

32. DI(K) is the correction to be applied to potential intake during lactation and suckling. These values may all be set to 1.0 initially as the corrections are calculated in subroutine STOCK.

33. EL(K) is the daily milk yield (MJ of metabolizable energy) from ewes in class K. Entered here (with a negative sign) only if the run starts during lactation.

34. PJ(I) is the priority value for each fodder unit status; at present these values run only from 1 to 4.

35. PK(I) is the priority value for each stock class status.

36. VJ(I) is the value of feed ($\$kg^{-1}$) for feed with digestibility 60% in unit with status I.

37. VK(I) is the value per head ($) of animals of normal body weight and over the age of 180 days in a class with status I. The value of other animals is obtained by calculation in subroutine OPTI (lines 40-41).

38. FI(I) are values which define the degree of rotational grazing to be imposed on a stock class with status I. For example, a value of 3.0 means that at least one-third of the feed supplies allocated to that class must be grazed at one time, i.e. a 3-paddock rotation, if enough paddocks are available. A value of 1.5 means that if 3 paddocks are available, the best 2 will be grazed, allowing the one remaining to be spelled. A value of 1.0 results in continuous grazing.

39. TI(I) is the length of time spent by a stock class in status I. For status 2 and 12, in the present example, a flushing period of 21 days is imposed, followed by a mating period of 34 days (status 3 and 13). Early pregnancy (status 4 and 14) is defined as the next 74 days, leaving 42 days for late pregnancy (status 5 and 15), while lambing (6 and 16) lasts another 34 days. The value of 160 days for status 1 is used to calculate the start of flushing at the same time in the following year (see subroutine REPRO).

40. TK(K) is the number of days from the start of each status at which the next stock event occurs. In this example, the fact that TK(1) = TI(1) indicates that no further stock event is scheduled until the status of class 1 ewes increases from 1 to 2.

41. Various parameters of pregnant ewes in classes 1 and 2. Values are entered here only if the run starts during gestation.
S0, S1, S2 are the proportions of ewes in each class with 0, 1 or 2 foetuses respectively.
B1, B2 are the weights of single and twin foetuses respectively (kg).
TMID is the interval (days) between the start of mating and

the mean conception date.
42. IOP(IO) is the reference number of the IOth optimization parameter. These numbers are specified as indices of the C array in common block C, as described under subroutine OPTI. In this example, the stocking rate is to be tested by changing the number of breeding ewes, C(81).
43. NOP is the number of parameters to be optimized.
44. DOP(I) is the initial step-size for each of the optimization parameters. These values are read in only at the start of the first run. For the single optimization parameter in this example, C(81), the initial value tested is 2500.
It is then increased by DOP(1) to 3000. If this results in no improvement, the reverse step size means that 2000 is tested. If all attempts to improve the objective function fail, the step size is reduced to 250, values of 2750 or 2250 are tested, and so on.

The pasture data file, LFG, contains a number of seasonal patterns of growth parameters in the form of the CFG array (see Section 3). Three rows of data are required for each year (ceiling yields, relative growth rates and senescence rates), with 27 values in each row (one for the first day and every 14 days thereafter. There is no limit on the number of years. The values used in this example and an illustration of the herbage yield in the absence of sheep are shown in Fig. 26.

Example of a single run lasting one year

The operation of the model over a single year, starting 1 January, is described, using the same parameters for herbage growth as for ungrazed pastures in Fig. 26. Values for the C array are shown in Table 16 and the main features of the chosen property are summarized in Tables 15a and b. The total area, A(20), is 170 ha (line 20) and is divided into only five paddocks (units 2-6) for the sake of simplicity, with area ratios of 5:4:3:2:1. At the start of the run, the amount of herbage dry matter on each paddock (line 19) is the same, 3000 kg ha^{-1}, and this consists mainly of dead material remaining after spring growth and senescence (lines 14 to 18). Of these fodder units, nos 5 and 6 are to be set aside for hay on or after day 220 if circumstances permit (line 12). Unit 1 (status 4) contains 200t hay with a mean digestibility 60% (C(19)) and 50t (C(20)) is to be purchased whenever the amount in the store falls to 50t.

The area is initially stocked with 2500 breeding ewes, four fifths of which consist of mature ewes in class 1, with an average age of 1000 days and body weight 55 kg, and one-fifth

U.C.W. ABERYSTWYTH LIBRARY

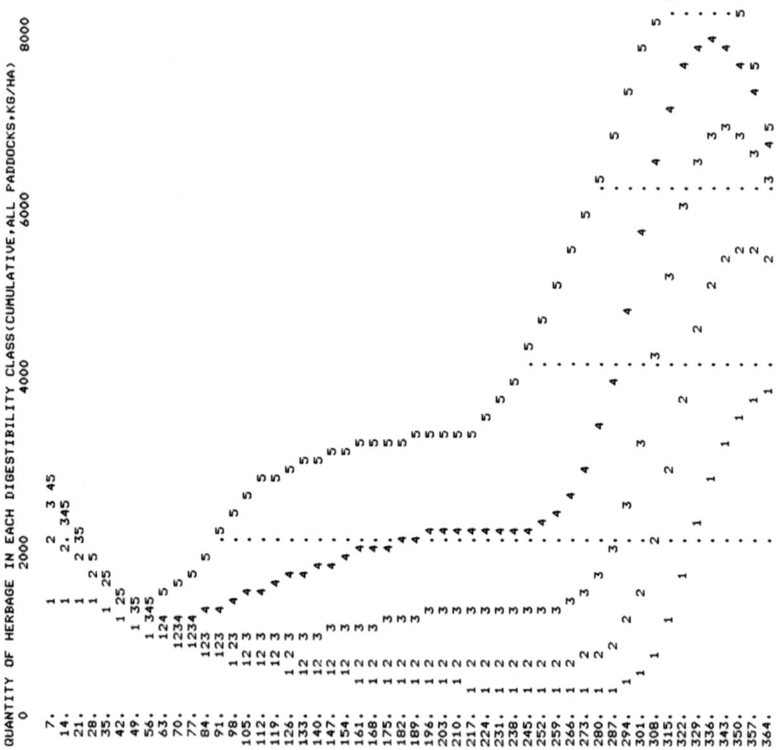

Ceiling yield (kg/ha)	Relative growth rate	Senescence factor
9200	0.005	4.0
9100	0.005	4.0
9100	0.010	2.0
9100	0.020	1.0
8900	0.025	0.5
8200	0.025	0.5
7200	0.025	0.5
6200	0.020	0.5
5600	0.010	0.5
5200	0.010	0.5
4900	0.005	0.5
4600	0.005	0.5
4400	0.005	0.5
4200	0.005	0.5
4200	0.005	0.5
4300	0.010	0.5
4600	0.020	0.5
5000	0.030	0.5
5500	0.040	0.5
6000	0.050	0.6
7000	0.050	1.0
8100	0.050	2.0
8600	0.040	3.0
8800	0.030	4.0
9100	0.020	4.0
9300	0.010	4.0
9200	0.005	4.0

Fig. 26. Mean weight of herbage on ungrazed paddocks, predicted from the indicated pattern of seasonal growth parameters. The symbols represent herbage classes and each point indicates a cumulative total including all less digestible classes.

of 17-month-old maiden ewes (class 2) weighing 50 kg (lines 27-30). In addition there are 500 unfinished weaners, 140 days old and weighing 30 kg, which are for sale (class 6). The initial weight of wool for each class is zero. Flushing starts for the mature ewes on day 30 and for the maiden ewes on day 50.

The priority values and grazing patterns adopted in this example (lines 34, 35 and 38) are based on general experience and preliminary testing, rather than systematic optimization searches and are therefore not necessarily regarded as the most suitable.

The allocation of stock classes to mobs at various times during the year is shown in Table 17. For the first 29 days, class 6 weaners are in mob 1 and all the other stock classes are grazed together in mob 2 which has a lower priority. On day 30, at the start of flushing, the maiden ewes are drafted into a third mob, leaving the mature ewes in mob 2. The maiden ewes move into mob 1 on day 50 when their priority increases. At the end of mating, a new class of non-pregnant ewes (class 3) is formed from the mature ewes (day 85) and this class is expanded from the maiden ewes on day 105. As these sheep have a lower priority than the pregnant ewes they graze in mob 3 with the ewe weaners which are purchased on day 100. The last of the class 6 weaners are sold on day 111 with the result that the pregnant ewes move into mob 1. By day 159 the mature ewes are six weeks from lambing and, having a higher priority, are left in mob 1 by themselves. They are rejoined by class 2 ewes 20 days later.

At weaning (day 308) the mature ewes are transferred to the lowest priority mob. When weaning occurs for the younger ewes (day 336), both classes of lambs for sale (classes 7 and 8) are given preferential treatment, with all other animals grazed together in mob 2.

No decision rules have been incorporated to keep incompatible classes separate, although such restrictions may apply in certain circumstances. In this example, no more than three mobs are ever formed, because the priority values fall into three fairly distinct categories. By specifying intermediate values for certain classes or a narrower range of values within one mob, a greater number of mobs would be produced.

The effects of these various re-allocations on the herbage availabilities are shown in Table 18. The high-priority mobs tend to be placed into paddocks of highest availability, while the paddocks of lowest availablity remain unoccupied; the apparent anomalies arise because relative intakes are also

Table 18. Herbage availability (F) and mob (MJ) on fodder units.

DRAFT DAY	UNIT STATUS 1	2	3	4	5	6	FEED (F(6,J)) AND MOB NO. (MJ) ON EACH UNIT 1 F	MJ	2 F	MJ	3 F	MJ	4 F	MJ	5 F	MJ	6 F	MJ
2.	4	1	1	1	1	1	200.	0	2935.	1	2935.	2	2935.	2	2935.	2	2935.	0
16.	4	1	1	1	1	1	200.	0	2220.	1	1939.	2	1939.	2	1939.	2	2352.	1
30.	4	1	1	1	1	1	200.	0	1639.	1	1221.	2	1221.	2	1221.	3	1745.	1
44.	4	1	1	1	1	1	200.	1	1243.	1	803.	3	803.	0	822.	2	1321.	1
50.	4	1	1	1	1	1	192.	1	1154.	1	744.	2	770.	2	581.	0	1223.	1
51.	4	1	1	1	1	1	190.	1	1142.	1	735.	2	761.	2	585.	0	1209.	1
65.	4	1	1	1	1	1	169.	0	1130.	1	718.	2	737.	2	778.	1	1180.	1
71.	4	1	1	1	1	1	169.	0	1159.	1	710.	2	728.	2	827.	1	1206.	1
85.	4	1	1	1	1	1	169.	0	1261.	1	673.	3	686.	2	963.	2	1305.	1
99.	4	1	1	1	1	1	169.	0	1361.	1	868.	2	583.	3	681.	2	1402.	1
100.	4	1	1	1	1	1	169.	0	1368.	1	848.	2	594.	3	678.	2	1409.	1
105.	4	1	1	1	1	1	169.	0	1394.	2	754.	2	582.	3	650.	3	1434.	1
112.	4	1	1	1	1	1	169.	0	1229.	1	691.	1	579.	2	637.	2	1397.	1
126.	4	1	1	1	1	1	169.	0	946.	1	584.	1	539.	2	580.	2	1044.	1
140.	4	1	1	1	1	1	169.	0	737.	1	511.	1	510.	2	540.	2	795.	1
154.	4	1	1	1	1	1	169.	0	593.	1	460.	1	490.	2	511.	2	628.	1
159.	4	1	1	1	1	1	169.	0	554.	2	446.	3	484.	1	503.	1	582.	1
173.	4	1	1	1	1	1	169.	1	561.	1	428.	3	392.	0	403.	2	450.	1
179.	4	1	1	1	1	1	161.	1	494.	1	422.	1	432.	1	392.	0	430.	2
193.	4	1	1	1	1	1	142.	1	452.	1	407.	2	408.	1	482.	1	268.	0
201.	4	1	1	1	1	1	129.	1	427.	1	402.	1	392.	2	426.	1	325.	0
215.	4	1	1	1	1	1	108.	1	406.	1	385.	2	368.	0	397.	1	425.	1
220.	4	1	1	1	1	1	100.	1	398.	0	401.	2	421.	1	388.	1	387.	1
221.	4	1	1	1	1	1	99.	1	410.	0	405.	1	409.	1	384.	2	381.	1
225.	4	1	1	1	1	1	93.	1	466.	1	397.	1	397.	1	381.	0	379.	2
234.	4	1	1	1	1	1	77.	0	477.	1	445.	1	443.	1	544.	1	305.	2
235.	4	1	1	1	1	1	77.	0	483.	2	453.	3	452.	2	541.	1	302.	0
248.	4	1	1	1	1	3	77.	0	494.	2	599.	1	476.	3	575.	2	648.	0
255.	4	1	1	1	1	3	77.	0	492.	2	758.	1	560.	1	514.	1	966.	0
262.	4	1	1	1	3	3	77.	0	673.	1	717.	1	574.	2	562.	0	1477.	0
276.	4	1	1	1	3	3	77.	0	749.	2	811.	1	999.	1	1559.	0	3222.	0
290.	4	1	1	1	3	3	77.	0	1676.	1	788.	2	987.	1	3554.	0	5304.	0
304.	4	1	1	1	3	3	77.	0	2558.	1	1647.	1	1226.	2	5841.	0	6913.	0
308.	4	1	1	1	3	3	77.	0	2859.	1	1808.	2	1430.	3	6319.	0	7205.	0
322.	4	1	1	1	3	3	77.	0	3996.	1	2855.	2	900.	3	7377.	0	7871.	0
324.	4	1	1	1	3	1	124.	0	4126.	1	2986.	2	813.	3	7468.	0	1000.	1
327.	4	1	1	1	1	1	214.	0	4304.	1	3165.	2	693.	3	1000.	1	1005.	3
336.	4	1	1	1	1	1	214.	0	4652.	1	3541.	2	521.	0	983.	0	721.	0
350.	4	1	1	1	1	1	214.	0	4531.	1	2552.	2	651.	2	975.	2	771.	0
355.	4	1	1	1	1	1	214.	0	4380.	1	2248.	2	597.	2	870.	2	761.	2

related to digestibilities, which are not shown here. The number of paddocks grazed by any mob also depends on the number of animals and the paddock area.

During flushing of the ewes there is little herbage growth, and the material on offer is of poor quality; consequently, hay is fed at this time. By day 173, at the end of autumn, the amount of herbage on all paddocks is very low. The additional requirements of the mature ewes in late pregnancy necessitates the further feeding of hay until day 259 (Table 19).

Table 19. Nutrition of mature ewes in class 1, in relation to intake of DM(kg), metabolizable energy (MJ) and body weight (kg)

NUTRITION OF SELECTED STOCK CLASS, K=1

DAY	PRIORITY	POTL. INTAKE	ACTUAL INTAKE	HAY INTAKE	INTAKE DDM	DIGEST- IBILITY	METAB. ENERGY	MILK ENERGY	ENERGY SURPLUS	WEIGHT GAIN	BODY WEIGHT
7.	-0.39	1.50	1.20	0.00	0.72	0.60	11.45	0.00	0.48	0.01	55.25
14.	-0.39	1.50	0.96	0.00	0.51	0.54	8.20	0.00	-0.90	-0.03	55.19
21.	-0.39	1.50	0.75	0.00	0.37	0.50	5.98	0.00	-1.90	-0.06	54.89
28.	-0.39	1.50	0.53	0.00	0.25	0.47	4.01	0.00	-2.75	-0.08	54.38
35.	-0.79	1.50	0.42	0.00	0.19	0.45	2.97	0.00	-3.18	-0.10	53.74
42.	-0.79	1.50	0.37	0.00	0.16	0.43	2.53	0.00	-3.34	-0.10	53.04
49.	-0.79	1.50	0.35	0.48	0.44	0.53	7.07	0.00	-1.19	-0.04	52.72
56.	-0.79	1.50	0.47	0.49	0.56	0.58	8.92	0.00	-0.29	-0.01	52.57
63.	-0.79	1.50	0.64	0.49	0.71	0.63	11.31	0.00	0.68	0.02	52.65
70.	-0.79	1.50	0.75	0.00	0.51	0.68	8.23	0.00	-0.42	-0.01	52.54
77.	-0.79	1.50	0.82	0.00	0.58	0.70	9.26	0.00	0.05	0.00	52.52
84.	-0.79	1.50	0.86	0.00	0.61	0.72	9.79	0.00	0.30	0.01	52.57
91.	-0.69	1.50	0.93	0.00	0.67	0.73	10.78	0.00	0.71	0.02	53.08
98.	-0.69	1.50	0.80	0.00	0.57	0.71	9.18	0.00	0.01	0.00	53.15
105.	-0.69	1.50	1.28	0.00	0.99	0.77	15.83	0.00	3.17	0.10	53.56
112.	-0.69	1.50	1.18	0.00	0.89	0.75	14.19	0.00	2.19	0.07	54.10
119.	-0.69	1.50	1.02	0.00	0.75	0.73	11.94	0.00	1.05	0.03	54.43
126.	-0.69	1.50	0.89	0.00	0.63	0.71	10.06	0.00	0.15	0.00	54.54
133.	-0.69	1.50	0.78	0.00	0.54	0.69	8.64	0.00	-0.54	-0.02	54.49
140.	-0.69	1.50	0.70	0.00	0.48	0.68	7.64	0.00	-1.06	-0.03	54.32
147.	-0.69	1.50	0.64	0.00	0.43	0.67	6.94	0.00	-1.45	-0.04	54.05
154.	-0.69	1.50	0.59	0.00	0.40	0.68	6.42	0.00	-1.76	-0.05	53.70
161.	-0.99	1.50	0.57	0.00	0.41	0.72	6.54	0.00	-1.75	-0.05	53.33
168.	-0.99	1.50	0.47	0.00	0.34	0.72	5.41	0.00	-2.35	-0.07	52.87
175.	-0.99	1.50	0.59	0.55	0.74	0.65	11.78	0.00	0.05	0.00	52.59
182.	-0.99	1.50	0.51	0.55	0.70	0.66	11.18	0.00	-0.41	-0.01	52.53
189.	-0.99	1.50	0.49	0.55	0.69	0.66	11.03	0.00	-0.73	-0.02	52.41
196.	-0.99	1.50	0.50	0.61	0.73	0.66	11.76	0.00	-0.72	-0.02	52.26
203.	-0.99	1.50	0.47	0.57	0.69	0.66	11.07	0.00	-1.27	-0.04	52.03
210.	-0.99	1.50	0.46	0.57	0.68	0.66	10.94	0.00	-1.62	-0.05	51.71
217.	-0.99	1.61	0.52	0.63	0.77	0.67	12.37	-5.66	-2.44	-0.08	51.29
224.	-1.00	1.91	0.70	0.68	0.95	0.69	15.15	-7.74	-2.68	-0.09	50.71
231.	-1.00	2.10	1.00	0.75	1.23	0.70	19.67	-8.27	-1.24	-0.04	50.32
238.	-0.80	2.17	1.18	0.00	0.93	0.79	14.83	-7.73	-2.50	-0.08	49.80
245.	-0.80	2.16	1.28	0.00	1.01	0.79	16.10	-6.44	-1.08	-0.04	49.42
252.	-0.80	2.12	1.33	0.00	1.05	0.79	16.79	-5.81	-0.37	-0.01	49.31
259.	-0.81	2.06	1.69	0.00	1.34	0.80	21.49	-5.34	2.03	0.07	49.63
266.	-0.81	1.99	1.77	0.00	1.41	0.80	22.56	-4.44	3.15	0.10	50.27
273.	-0.81	1.93	1.74	0.00	1.39	0.80	22.18	-3.64	3.55	0.11	51.04
280.	-0.81	1.86	1.75	0.00	1.40	0.80	22.33	-2.95	4.09	0.13	51.94
287.	-0.81	1.81	1.64	0.00	1.30	0.79	20.79	-2.36	3.76	0.12	52.80
294.	-0.81	1.75	1.73	0.00	1.38	0.80	22.02	-1.88	4.68	0.14	53.75
301.	-0.81	1.71	1.69	0.00	1.34	0.80	21.48	-1.48	4.67	0.14	54.75
308.	-0.39	1.67	1.65	0.00	1.30	0.79	20.87	0.00	5.47	0.16	55.77
315.	-0.39	1.50	1.36	0.00	1.00	0.73	15.94	0.00	2.81	0.08	56.51
322.	-0.39	1.50	1.06	0.00	0.71	0.67	11.35	0.00	0.69	0.02	56.84
329.	-0.39	1.50	0.72	0.00	0.43	0.60	6.93	0.00	-1.25	-0.04	56.74
336.	-0.39	1.50	1.50	0.00	1.19	0.79	19.05	0.00	4.97	0.14	56.60
343.	-0.39	1.50	1.46	0.00	1.06	0.73	16.94	0.00	3.16	0.09	57.37
350.	-0.39	1.50	1.07	0.00	0.67	0.63	10.70	0.00	0.27	0.01	57.77
357.	-0.39	1.50	0.89	0.00	0.51	0.58	8.19	0.00	-1.02	-0.03	57.79
364.	-0.39	1.50	0.73	0.00	0.38	0.53	6.14	0.00	-1.97	-0.06	57.48

During the winter months, herbage production barely keeps up with consumption and it is not until days 262 and 248 that sufficient feed is present to allow fodder units 5 and 6 respectively to be closed for hay. No paddocks were closed to defer grazing because of the generally unfavourable season; at no time during the first half of the year was the feed supply in excess of requirement.
The small number of paddocks used in this example restricted the flexibility of management, and more paddocks, combined with more frequent feed evaluations, should provide more effective control, as well as eliminating some of the abrupt changes in nutrition experienced here.
The mean amounts of herbage in each digestibility class, taken over the whole farm area, are shown in Fig. 27. The much lower availability of all classes than for the same area when ungrazed is evident throughout the year. The amount of feed present at the end of the year is slightly less than at the beginning, but any difference disappears during the first few weeks of the year.

Body weight changes in the different classes during the year are shown in Fig. 28. The general trends follow herbage availabilities, falling in late summer, rising with autumn growth, declining steadily during winter, and increasing rapidly during the spring. Preferential mob treatment resulted in small but consistent benefits, although the effect is sometimes obscured by other factors. For instance, during the period from day 159 to day 213, the dry ewes in mob 2 lost less weight than the pregnant mature ewes in mob 1, although the class 1 ewes had a much higher priority at this time. The difference in nutrition was considerable but this was used by the pregnant ewes for the development of the conceptus (which is excluded from the body weight of the ewe).

Towards the end of the year the picture is somewhat complicated by the restructuring of classes (day 355) and the fortnightly sale of heavier weight lambs in classes 6-8. Nevertheless, the mean weight of each stock class at the end of the year is comparable with that at the beginning.

Stock numbers at the times of changes in class status are shown in Table 20. Conception rates were 124% and 129% in the mature and maiden ewes respectively. At the restructuring of classes on day 355, a total of 416 ewes was culled from classes 1 to 3, restoring the number in class 2 to the original value. In addition to the breeding flock is the class of 405 lambs that are still below market weight at the end of the year.

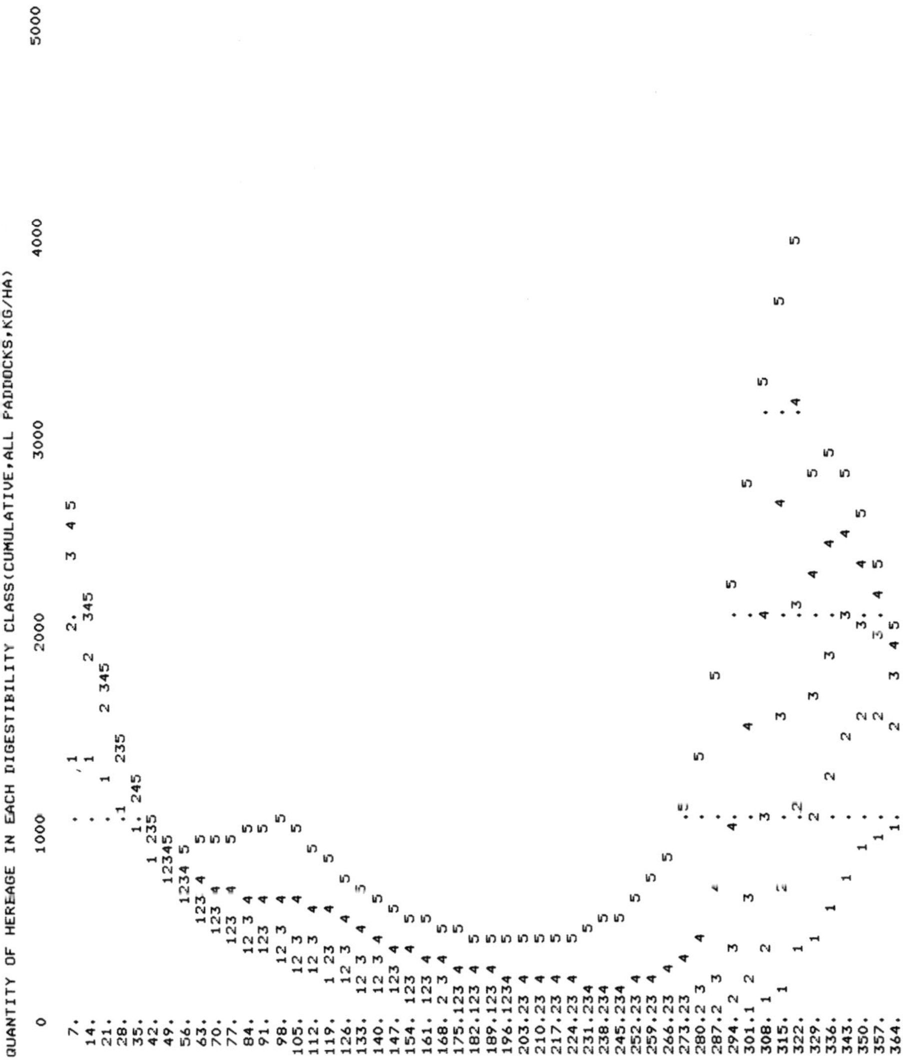

Fig. 27. Mean predicted weight of herbage on grazed paddocks, using the same growth parameters as in Fig. 26. The symbols represent herbage classes and each point indicates a cumulative total including all less digestible classes.

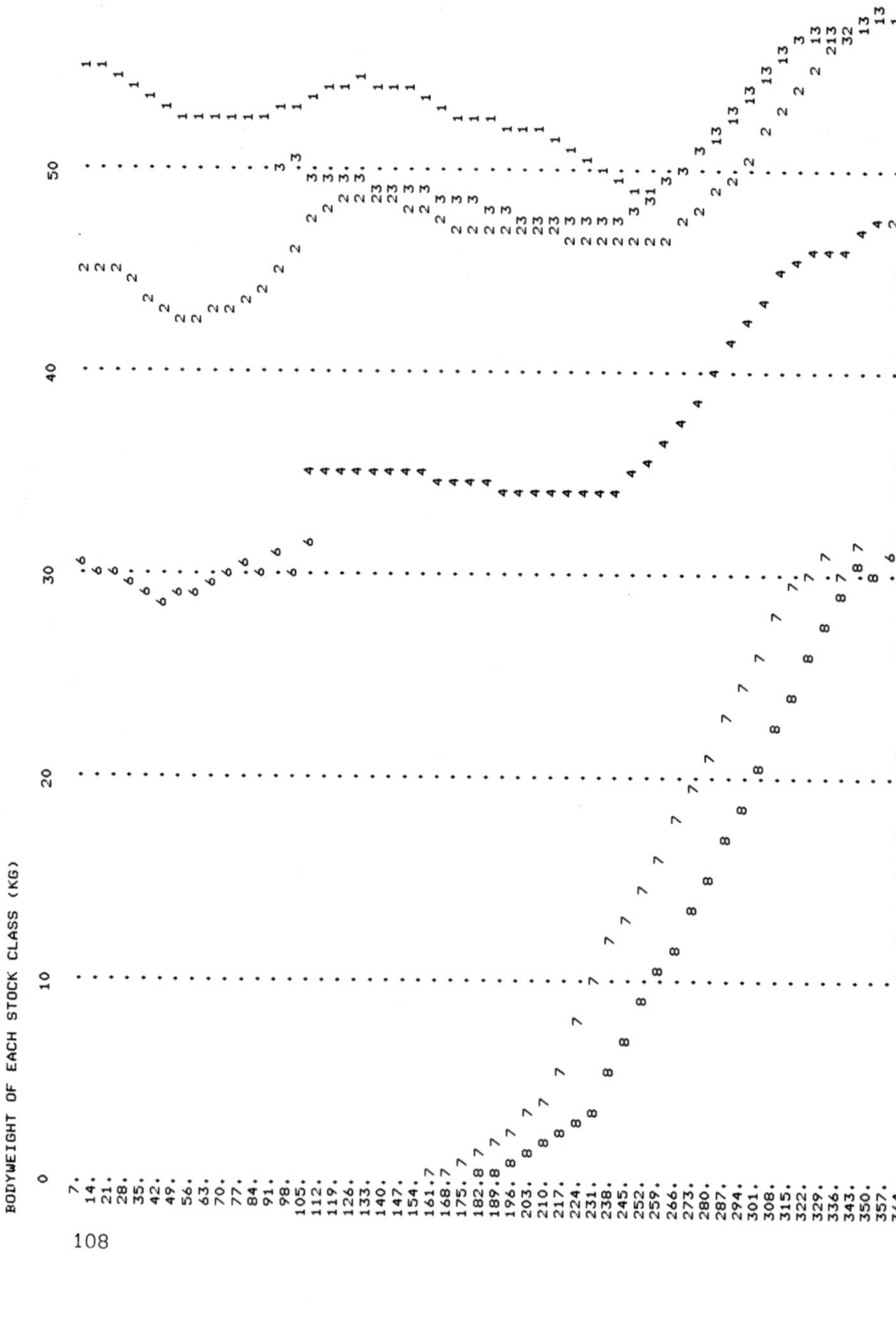
BODYWEIGHT OF EACH STOCK CLASS (KG)
0
10
20
30
40
50

Table 20. Stock numbers at status changes.

DAY	CLASS	STATUS	AGE	SHEEP NUMBER 1	2	3	4	5	6	7	8
14.	6	9	153.	1997.	499.	0.	0.	0.	419.	0.	0.
28.	6	9	167.	1995.	499.	0.	0.	0.	418.	0.	0.
30.	1	2	1029.	1994.	499.	0.	0.	0.	418.	0.	0.
42.	6	9	181.	1992.	498.	0.	0.	0.	418.	0.	0.
43.	1	2	1042.	1992.	498.	0.	0.	0.	418.	0.	0.
50.	2	12	554.	1990.	498.	0.	0.	0.	417.	0.	0.
51.	1	3	1050.	1990.	498.	0.	0.	0.	417.	0.	0.
56.	6	9	195.	1989.	497.	0.	0.	0.	417.	0.	0.
60.	1	3	1059.	1988.	497.	0.	0.	0.	417.	0.	0.
62.	2	12	567.	1988.	497.	0.	0.	0.	417.	0.	0.
70.	6	9	209.	1986.	497.	0.	0.	0.	412.	0.	0.
71.	2	13	575.	1986.	497.	0.	0.	0.	412.	0.	0.
77.	1	3	1076.	1985.	496.	0.	0.	0.	412.	0.	0.
80.	2	13	584.	1984.	496.	0.	0.	0.	412.	0.	0.
84.	6	9	223.	1983.	496.	0.	0.	0.	302.	0.	0.
85.	1	4	1084.	1805.	496.	178.	0.	0.	302.	2450.	0.
97.	2	13	601.	1803.	495.	178.	0.	0.	301.	2447.	0.
98.	6	9	237.	1802.	495.	178.	0.	0.	158.	2447.	0.
100.	4	10	240.	1802.	495.	178.	525.	0.	158.	2446.	0.
105.	2	14	609.	1801.	455.	218.	524.	0.	158.	2445.	619.
112.	6	9	0.	1800.	455.	217.	524.	0.	0.	2444.	618.
159.	1	5	1158.	1792.	453.	216.	522.	0.	0.	2432.	616.
171.	1	5	1170.	1789.	452.	216.	521.	0.	0.	2429.	615.
179.	2	15	683.	1788.	452.	216.	521.	0.	0.	2427.	614.
192.	2	15	696.	1786.	451.	216.	520.	0.	0.	2424.	613.
201.	1	6	1200.	1784.	451.	215.	519.	0.	0.	2422.	613.
213.	1	6	1212.	1774.	451.	215.	519.	0.	0.	2045.	612.
221.	2	16	725.	1773.	450.	215.	518.	0.	0.	2043.	612.
225.	4	11	365.	1772.	450.	215.	518.	0.	0.	2042.	611.
234.	2	16	738.	1770.	449.	215.	518.	0.	0.	2040.	475.
235.	1	7	1234.	1770.	449.	215.	518.	0.	0.	2040.	475.
255.	2	17	759.	1767.	448.	214.	517.	0.	0.	2036.	474.
294.	1	7	1293.	1760.	447.	213.	515.	0.	0.	2028.	472.
308.	1	1	1307.	1757.	446.	213.	514.	0.	0.	2025.	472.
322.	2	17	826.	1755.	445.	213.	513.	0.	0.	2022.	471.
322.	7	9	109.	1755.	445.	213.	513.	0.	0.	1672.	471.
336.	2	1	840.	1752.	445.	213.	512.	0.	0.	1670.	470.
336.	7	9	123.	1752.	445.	213.	512.	0.	0.	890.	470.
350.	8	9	116.	1750.	444.	212.	512.	0.	0.	888.	291.
350.	7	9	137.	1750.	444.	212.	512.	0.	0.	364.	291.
355.	1	1	990.	1989.	511.	0.	0.	0.	655.	0.	0.
364.	6	9	142.	1987.	511.	0.	0.	0.	409.	0.	0.

Further information on the nutrition of the mature ewe class is provided in Table 19. The potential intake DK(1) is shown in column 3, and the extent of the rise which occurs during lactation and reaches a peak between days 238 and 245 is indicated. The actual dry matter intake DF(column 4) falls to its lowest values in late summer and winter, and rises to approach the potential intake between days 294 and 308. During flushing and late pregnancy, the amount of hay dry matter fed, DHAY, is about 0.5 kg day^{-1} (column 5). The digestible dry matter intake of herbage and hay combined and the mean digestibility are shown in columns 6 and 7.

The changes in the intake of metabolizable energy (column 8) reflect the extremes in nutrition which are imposed during the year. The energy expenditure in milk production (column 9) exhibits a peak between 224 and 238 days, slightly before the corresponding peak in potential intake and well before the peak in actual energy intake. The net excess in energy intake and consequent daily body weight gains are shown in columns 10 and 11.
The overall changes during the year are brought together in the objective function (Table 21). Hay was made on fodder units 5 and 6 but the mean weight of herbage on the paddocks at the end of the year is similar to the starting value. The difference in the overall value of fodder is relatively small and the condition of the sheep has been maintained.

A total of 2652 lambs, valued at $27,731, were sold, while the 2092 animals shorn at the end of the year produced 10095 kg of clean wool, averaging 3.84 kg per head for the breeding flock. When variable costs per ewe are subtracted, the increase in the objective function over the year is $168/ha. This estimate of course depends entirely upon prevailing prices for animals and hay and upon the assessments of the worth of the various categories of stock and standing herbage on the property at a given time.
The pattern of body weight changes predicted from this run is similar to results obtained, during similar seasons, from a long-term experiment at the Ginninderra Experiment Station, near Canberra, using the same type of sheep grazing at the same stocking rate.

Changes to the grazing system

Reference was made at the beginning of this Section to the structural modifications that the operator should make if different classes of animals are to be included in the model. However, even smaller changes to the system will often require a number of adjustments to the data arrays. For example, a major change in the specified data for mating might mean that the run would start on 1 January with pregnant or lactating ewes. The operator can choose between starting the run on a different calendar date or changing the initial status values of the stock classes.

If the first course is adopted, the value of C(1) becomes the calendar date at the start so that the CFG array in the LFG file will be read from the correct point; it is also important with regard to this file that the initial and final values in each line of the CFG array should be the same. The initial weight of herbage in each digestibility class in each fodder unit must be adjusted to suit the new starting date. Changes should also be made to the data relevant to haymaking: the times, T(J,2) for considering a change in fodder unit status and the limiting dates for haymaking, C(12) and C(13). However, it is often preferable to maintain the starting date at a time of year when the pastures are dormant. In this case the operator should adjust the initial stock class parameters accordingly.

Table 21a. Objective function at start of year.

RUN 1 YEAR 1 FOD.GROWTH 2

TOTAL FEED/HA. ON EACH UNIT								
200.	3000.	3000.	3000.	3000.	3000.	3000.	3000.	3000.0
STOCK NUMBER IN EACH CLASS								
2000.	500.	0.	0.	0.	500.	0.	0.	3000.0
STOCK BODYWEIGHTS								
55.0	45.0	0.0	0.0	0.0	30.0	0.0	0.0	0.0
WOOL WEIGHTS								
0.00	0.00	0.00	0.00	0.00	0.00	0.00	0.00	0.0
TOTAL VALUE OF FEED ON FODDER UNITS								
8000.	2168.	1734.	1301.	867.	434.	0.	0.	85.3
TOTAL VALUE OF STOCK IN EACH CLASS								
39975.	9955.	0.	0.	0.	1936.	0.	0.	305.1
VALUE OF WOOL FOR EACH STOCK CLASS								
0.	0.	0.	0.	0.	0.	0.	0.	0.0
VARIABLE COSTS FOR EACH STOCK CLASS								
0.	0.	0.	0.	0.	0.	0.	0.	0.0

Table 21b. Objective function at end of year.

```
OBJ.F FOR THIS RUN        0.00     PREVIOUS BEST OBJ.F.    -999.00

VALUES OF OPTIMIZATION PARAMETERS

2500.0

INCOME-EXPENSES FOR FODDER UNITS
    0.     0.     0.     0. -896. -471.     0.     0.          -8.0

INCOME-EXPENSES FOR STOCK CLASSES
 2082.     0.     0.     0.     0.     0.     0.     0.     0.  124.6

 TOTAL FEED/HA. ON EACH UNIT
   214. 3759. 1595.   479.   658.   579. 3000. 3000.           65.2

STOCK NUMBER IN EACH CLASS
 1986.   511.     0.     0.     0.   408.     0.     0.        2905.7

STOCK BODYWEIGHTS
  57.4  47.3   0.0   0.0   0.0  30.2   0.0   0.0               0.0

WOOL WEIGHTS
  3.95  3.40  0.00  0.00  0.00  1.20  0.00  0.00               0.0


TOTAL VALUE OF FEED ON FODDER UNITS
 9036. 2178.   594.   229.   158.    80.     0.     0.          72.2

TOTAL VALUE OF STOCK IN EACH CLASS
39719.10163.      0.     0.     0. 1610.     0.     0.         302.9

VALUE OF WOOL FOR EACH STOCK CLASS
15688. 3478.      0.     0.     0.   984.     0.     0.        118.5

VARIABLE COSTS FOR EACH STOCK CLASS
-9620.     0.     0.     0.     0.     0.     0.     0.         -56.6

OBJ.F FOR THIS RUN     163.17  PREVIOUS BEST OBJ.F.    -999.00

VALUES OF OPTIMIZATION PARAMETERS

2500.
```

Acknowledgement

The authors acknowledge the financial assistance of the Australian Wool Corporation in this work.

References

A.R.C., 1965. Nutrient Requirements of Farm Livestock No. 2-Ruminants. Agric. Res. Counc.: London.

Alexander, G., I. McCance, and R.H. Watson, 1955. Aust. vet. J. 31: 85-90.

Allden, W.G., 1968. Aust. J. agric. Res. 19: 997-1007.

Arnold, G.W., 1964. In: Grazing in Terrestrial and Marine Environments. p. 133-154. Butterworths, London.

Arnold, G.W. and M.L. Dudzinski, 1967. Aust. J. agric. Res. 18: 349-359.

Balch, C.C., 1971. Br. J. Nutr. 26: 383-392.

Blaxter, K.L., 1968. In: Growth and Development of Mammals. G.A. Lodge and G.E. Lamming (Eds), p. 329-344. Butterworths, London.

Brody, S., 1945. Bioenergetics and Growth. Hafner, New York.

Christian, K.R., J.S. Armstrong, J.L. Davidson, J.R. Donnelly and M. Freer, 1976. Proc. 12th. Int. Grassld Congr., Moscow, 1974. 3: 105-110.

Christian, K.R., J.S. Armstrong, J.R. Donnelly, J.L. Davidson and M. Freer, 1972. Proc. Aust. Soc. Anim. Prod. 9: 124-129.

Corbett, J.L., 1968. Aust. J. agric. Res. 19: 283-294.

Curll, M.L., J.L. Davidson and M. Freer, 1975. Aust. J. agric. Res. 26: 553-565.

Davies, H.L., 1963. Aust. J. agric. Res. 14: 824-838.

Donnelly, J.R., J.L. Davidson and M. Freer, 1974. Aust. J. agric. Res. 25: 813-823.

Ferguson, K.A., 1962. Aust. J. biol. Sci. 15: 720-731.

Ferguson, K.A., 1970. In: Feeding Protected Protein to Sheep and Cattle. Proceedings Aust. Soc. Anim. Prod. (N.S.W. Branch) Sept. 1970. D.W. Horwood (Ed.)

Freer, M., J.L. Davidson, J.S. Armstrong and J.R. Donnelly, 1970. Proc. 11th. Int. Grassld Congr., Surfers Paradise, p. 913-917.

Graham, N.M., 1961. Publs Eur. Ass. Anim. Prod. 10: 226-

Graham, N.M., 1964a. Aust. J. agric. Res. 15: 127-41.

Graham, N.M., 1964b. Proc. Aust. Soc. Anim. Prod. 5: 272-79.

Graham, N.M., T.W. Searle and D.A. Griffiths, 1974. Aust. J. agric. Res. 25: 957-71.

Hadjipieris, G. and W. Holmes, 1966. J. agric. Sci., Camb. 66: 217-223.

Killeen, I.E., 1967. Aust. J. exp. Agric. Anim. Husb. 7: 126-136.
Langlands, J.P. and H.A.M. Sutherland, 1968. Br. J. Nutr. 22: 217-27
McKinney, G.T., A. Axelsen and F.H.W. Morley, 1970. Proc. Aust. Soc. Anim. Prod. 8: 466-71.
Obst, J.M. and H.R. Day, 1968. Proc. Aust. Soc. Anim. Prod. 7: 239-42.
Radford, H.M., 1959. Aust. J. agric. Res. 10: 377-86.
Rattray, P.V., W.N. Garrett, N.E. East and N. Hinman, 1974. J. Anim. Sci. 38: 613-26.
Reid, R.L. and N.T. Hinks, 1962. Aust. J. agric. Res. 13: 1092-1111.
Russel, A.J.F., J.M. Doney and R.L. Reid, 1967. J. agric. Sci., Camb. 68: 351-58.
Schinckel, P.G. and B.F. Short, 1961. Aust. J. agric. Res. 12: 176-202.
Searle, T.W. and N.M. Graham, 1970. Proc. Aust. Soc. Anim. Prod. 8: 472-75.
Van Dyne, G.M. and Z. Abramsky, 1975. In: Study of Agricultural Systems. G.E. Dalton (Ed.). Applied Science, London.
Walker, D.M. and B.W. Norton, 1971. J. agric. Sci., Camb. 77: 363-69.
Watson, R.H. and H.M. Radford, 1966. Aust. J. agric. Res. 17: 335-45.
White, D.H., 1975. Ph.D. Thesis. Univ. of New South Wales.
Willoughby, W.M., 1959. Aust. J. agric. Res. 10: 248-68.
Wood, P.D.P., 1969. Anim. Prod. 11: 307-316.

U.C.W. ABERYSTWYTH
LIBRARY